Acknowledgements

I would like to thank, firstly, Sir Hayden Phillips, Permanent Secretary and Clerk of the Crown and Chancery; Paul Monaghan and Jon Wallsgrove at the Court Service of the Lord Chancellor's Department; and Gordon Beaton and Ronald Gardiner of the Scottish Court Service for their kind cooperation without which this project would not been possible; Dr Clare Graham for her invaluable contribution, advice and vast knowledge of the subject; Allan Brodie, Garry Winter, Will Holbrough and Stephen Porter at English Heritage for help and information based on their own research; Rita Harkin and Harriet Devlin of the Ulster Architectural Heritage Society; Martin Charles for his wonderful photographs; all the staff of the many law courts we visited for their kindness and assistance; the many conservation officers who responded to enquiries for information. Amongst the many other individuals that offered advice, thoughts and information I would like in particular to thank Jim Stevenson, Terry Pawson, Alison Murdoch, Roger Evans, Geoffrey Goldsmith, Peter Barnes, Ann Bond, Eric Langham, Mr and Mrs Leo Austin, Nick Ludlow and Walter Boyd; and finally, and most importantly, at SAVE, Emily Cole for invaluable research, Deborah Churchill, Christopher Sperling, Adam Wilkinson and Marcus Binney for everything else.

Front cover:
The redundant Georgian courtroom in St Albans Town Hall
© Martin Charles

Inside front cover:
In Keswick Magistrates' Court pristine cells retain their beautiful iron doors and door furniture
© Martin Charles

Back cover:
Linenfold panelling and chairs, Boston Sessions House
© Martin Charles

Photographic Credits
Unless otherwise acknowledged, images are the copyright of SAVE Britain's Heritage.

Silence in Court: The Future of the UK's Historic Law Courts is published by SAVE Britain's Heritage, a registered charity (No. 269129) founded in 1975 to campaign for historic buildings at risk. Information about other SAVE publications and the organisation's current campaigns and activities can be found at:
www.savebritainsheritage.org

SAVE Britain's Heritage
70 Cowcross Street
London EC1M 6EJ
020 7253 3500
email: save@btinternet.com

Published February 2004

© SAVE Britain's Heritage
ISBN 0-905978-43-9

Fine panelling in Liverpool County Sessions House.
The building is currently used for storage and its future is uncertain.

This publication has been sponsored by:
English Heritage
Cadw – Welsh Historic Monuments
David Cooper & Co. Solicitors
Jones Day Gouldens

© Ray Main, Mainstream Photography

Contents

1. **Introduction** — 1
   An unexplored treasure trove
   by Marcus Binney

2. **Refurbishment, redundancy and reuse: the issues** — 4
   - Reform and modernisation: the causes of redundancy — 7
   - Northern Ireland: politicised buildings — 20
   - Scotland: a different path — 23
   - Redundancy: preservation or reuse? — 26
   - Conclusion — 34

3. **A history of law court architecture in England and Wales** — 36
   The institutionalisation of the law
   by Clare Graham

4. **It can be done: modernising historic law courts – an architect's perspective** — 48
   by Jim Stevenson

5. **Gazetteer**

   England and Wales
   - Appleby Shire Hall — 56
   - Barnard Castle Magistrates' Court — 57
   - Bedford Shire Hall — 58
   - Beverley Guildhall — 60
   - Beverley Sessions House — 62
   - Birkenhead Sessions House — 63
   - Birmingham Victoria Law Courts — 64
   - Bodmin Shire Hall — 66
   - Bolton County Court — 68
   - Bolton Magistrates' Court — 69
   - Boston Sessions House — 70
   - Bridgwater County Court — 72
   - Canterbury Sessions House — 73
   - Carlisle Assize Courts — 74
   - Cheltenham County Court — 76
   - Chester Shire Hall — 78
   - Chesterfield Magistrates' Court — 79
   - Chipping Camden Magistrates' Court — 80
   - Derby Shire Hall — 81
   - Devizes Assize Court — 84
   - Downham Market County Court — 85
   - Durham Crown Court — 86
   - Ely Shire Hall — 88
   - Exeter Sessions House — 90
   - Grantham Magistrates' Court — 91
   - Huntingdon Town Hall — 92
   - Kendal Town Hall — 93
   - Keswick Magistrates' Court — 95
   - Knutsford Sessions House — 96
   - Lancaster Shire Hall — 97
   - Leicester Castle — 98
   - Lincoln County Hall — 100
   - Liverpool County Sessions House — 101
   - Liverpool St George's Hall — 102
   - London Magistrates' Courts — 104
   - Lutterworth Magistrates' Court — 110
   - Maldon County Court — 111
   - Manchester Minshull Street Crown Court — 112
   - Manchester Magistrates' Court — 114
   - Morpeth Sessions House — 116
   - Northampton Sessions House — 118
   - Nottingham Shire Hall — 120
   - Oakham Castle — 121
   - Oldham County Court — 123
   - Oxford County Hall — 124
   - Presteigne Shire Hall — 126
   - Ripon Court House — 129
   - St Albans Town Hall and Courthouse — 130
   - Salford County Court — 132
   - Salford Magistrates' Court — 133
   - Salisbury Guildhall — 134
   - Sheffield County Court — 136
   - Sheffield Courthouse — 137
   - South Molton Guildhall — 138
   - Spalding Sessions House — 140
   - Spilsby Sessions House — 141
   - Wakefield Courthouse — 142
   - Walsall County Court — 143
   - Warrington County Court — 144
   - Warwick Shire Hall — 145
   - Watford County Court — 146
   - Wigton Magistrates' Court — 147
   - Windermere Magistrates' Court — 148
   - York Assize Court — 149

   Scotland
   - Dumbarton Sheriff Court — 151
   - Dundee Sheriff Court — 152
   - The Supreme Courts of Scotland, Edinburgh — 153
   - Forfar Sheriff Court — 155
   - Greenlaw Town Hall — 156
   - Perth Sheriff Court — 157
   - Stirling Sheriff Court — 158

   Northern Ireland
   - Armagh Courthouse — 159
   - Armagh Manor Courthouse — 162
   - Belfast Crumlin Road Courthouse — 163
   - Bushmills Courthouse — 165
   - Caledon Courthouse — 166
   - Hillsborough Courthouse — 167
   - Londonderry Courthouse — 169
   - Middleton Markethouse — 170

Introduction

An unexpected treasure trove
Marcus Binney

BEGINNING with railway architecture in 1977, SAVE has published a major series of campaigning studies of endangered, forgotten and neglected building types. These include Pennine textile mills (*Satanic Mills*), historic theatres (*Curtains!!!*), northern nonconformist chapels (*The Fall of Zion*), naval and military enclaves (*Deserted Bastions*) and the great series of mental asylums built by counties and county boroughs during the nineteenth and early twentieth century (*Mind over Matter*).

Now it is the turn of historic courts. They are the UK's great undiscovered architectural treasure-trove. Throughout the country there are courtrooms large and small, in medieval castles and guildhalls, in seventeenth century sessions

St. George's Hall, Liverpool. The magnificent Crown Court

houses, eighteenth century shire halls and assize courts and in mighty Victorian and Edwardian town halls, full of fine plasterwork, wonderful woodwork and splendid staircases. Yet they have remained remarkably little studied and rarely recorded. Law courts, though public buildings and familiar landmarks in both cities and county towns, rarely occur in histories of architecture except where they are by well-known architects. Examples are the Shire Hall at Warwick designed by the artful Sanderson Miller, the Assize Court at York by the Palladian architect John Carr and the splendid courts in Chester and Lancaster Castles by Thomas Harrison.

The special quality of historic courts lies in grand classical (and occasionally gothic) architecture, but above all in well fitted out interiors. The best are as complete and perfectly-fitted out examples of the joiner's art as an eighteenth century box-pewed church. Yet while the nineteenth century ecclesiologists stripped out or cut down almost every box pew, the conservatism of the legal profession ensured that many courtrooms survive unaltered from earlier centuries. There are columned and domed interiors to compare to the finest City churches, interiors with seating as complete as in the finest surviving Georgian theatres.

Quarter sessions and assizes in county towns were long a focus of local life, but an absolute ban on the photography of trials, enshrined in the Courts Act of 1927, has been widely interpreted as a prohibition of photography of all kinds in court buildings. The result is that some of Britain's most splendid interiors have been rarely photographed, let alone published. Michael Heseltine, when Secretary of State for the Environment in the early 1980s, commissioned an impressive three volume illustrated register of historic buildings in Government ownership, including all the courts operated by the Lord Chancellor's Department. Lamentably his civil servants short-changed him, doing little more than publish photographs of facades and quotes from Pevsner's county guides, failing in most cases to mention, let alone illustrate, important or interesting interiors. Scotland did far better with a first class volume, Historic Buildings at Work, prepared by the Scottish Civic Trust, which documents the interiors of many fine Sheriff Courts.

Although British justice has long been held up as a model for the world – habeas corpus, the jury system, the principle of innocence until guilt is proven – the Law has been as careless of its heritage as any great institution in Britain.

Even before SAVE embarked on this report we knew some of the best courts in the country were standing empty – the two great courtrooms in St George's Hall in Liverpool, arguably the grandest Neoclassical building in Europe; the seventeenth century Assize Courts in Northampton with their marvellous plasterwork; Sir Robert Smirke's courtrooms in the Citadel at Carlisle. The situation was largely caused by a £500 million court rebuilding programme undertaken by the Lord Chancellor's Department. This has been followed by an even more wide-ranging programme for rebuilding and regrouping Magistrates' Courts. It was astounding to visit the Arts and Crafts Magistrates' Courts in Birmingham, the masterpiece of Aston Webb, and find that closure was actually a possibility under the Government's Public Finance Initiative process.

Old courtrooms, it is sometimes argued, are unsuitable and impractical for modern use, lacking meeting rooms for barristers and clients and separate access for witnesses and jury members. Yet looking at new courtrooms, the most striking thing is usually how like the old courtrooms they look. As this report shows a growing number of courts are now being sympathetically modernised rather than rebuilt. In Preston the magnificent Edwardian Sessions House looked doomed when new Combined Courts opened in the 1980s, but rising demand for court space led the Lord Chancellor's Department to refurbish the old building for continued use. The furniture in the courtrooms was reversed so that witnesses come in from the same side as the judges and share their privacy. The old courtrooms have the advantage of much larger public galleries, taking 40 people rather than 20. Seats are separated by a central aisle so that in a tense murder trial, the victim's family can be seated on one side and the defendant's on the other.

Redundant courts pose a problem. Much of the interest lies in the woodwork and the tightly configured arrangement of bench, dock, witness box, seating for jurors and lawyers and public gallery. Where no more than the magistrates' bench and the witness box remains, the character of the place is lost. Yet Presteigne, the

Presteigne Shire Hall 1826-9.
The superbly restored dining room in the Judge's Lodgings. Note the remarkable survival of the mid-nineteenth century dinner service and other furnishings.

county town of Radnor in Wales, has set a model in preserving its courthouse complete with its fully-furnished judge's lodgings and opening it to the public. Bodmin has followed. There is a constant demand for court rooms for filming both television programmes and feature films.

As reorganisation of the UK's courts continues, more and more towns and cities will be faced with the closure of landmark buildings and the problems of finding suitable alternative uses. Richard Pollard's report provides the all-important evidence that Britain's law courts constitute one of the most remarkable and best preserved groups of public buildings in the country. It also provides vital ammunition for any community seeking to maintain a local group of courts in use or to find a new use for them. Both have been done successfully. Recently campaigners in Knutsford succeeded in overturning the Court Service's decision to close their Georgian Crown Court; now funds have been found to refurbish it. A similar alliance of press, public, lawyers, councillors and MPs has this spring triumphed in their battle to keep open the Grade II listed Magistrates' Court in Kingston upon Thames. These stories – and others – give hope to other communities fighting closures that the Lord Chancellor's Department (or its successor) can be made to bend to local wishes if resistance is widespread, organised and determined.

It is strange that the first study of this important subject is written and published not by Government, nor by one of the many arms of the Law, nor by University academics but by a voluntary preservation group with the slenderest of resources. SAVE's work in finding solutions for endangered buildings has always depended on efforts both at national and local level, and given growing local concern for the future of individual historic courts, no more fine court buildings need to be left empty and decaying, stripped of their fine interiors, let alone demolished. It is a large task, but a manageable and achievable one.

Refurbishment, redundancy and reuse: the issues

LAW courts are well known, much loved parts of the fabric of our towns and cities – elegant backdrops to our daily lives. And yet very few of us have ever thought to step inside a real law court, or had reason to. 70% of us only go to court once in our lives; less than half a percent of the population of West Derbyshire visit a court in any one year, according to a recent study. For most of us they are intimidating places full of unfamiliar and arcane ritual and procedure, the domain of the professional insiders – the lawyers – and criminals.

In one sense their interiors are very familiar: we all have a very firm idea of what a courtroom should look like thanks to film and TV's love affair with courtroom drama. Yet few of us know much about the architecture of individual court interiors, and quite how wonderful many of them are, thanks to the ban on the photography of trials enshrined in the 1927 Courts Act being interpreted as a prohibition of photography of any kind inside law courts. And because, as a result, law courts are such an under-appreciated part of our heritage, few are aware that historic law courts are an endangered species, that the massive and continuing programmes of legal and courts accommodation reforms have had an unprecedented impact on these buildings.

In the past thirty years hundreds of courts have closed, and many of the survivors will follow in the next thirty. Not all of these were great, or even good buildings, and many

The Royal Courts of Justice, London.
G.E. Street's masterpiece opened in 1882, the eventual outcome of a bitterly political competition. The great vaulted central hall – Street was determined to build the first stone-vaulted hall in England for centuries – is the heart of the building, alive with barristers, reporters and supporters hurrying to and from cases.

Magistrates' Courts in particular consisted of little more than rooms in pubs or the part-time use of town and village halls. But many very fine courthouses and courtrooms have been closed, and many of these remain empty today.

Redundancy poses potential problems for all historic buildings, but a particular threat to law courts because they are highly specialized, with unusual and complicated plans that make

The Old Bailey, London.
Perhaps no courthouse is better known than the Old Bailey, more properly called the Central Criminal Court. But because of the 1927 bar on trial photography, artists sketches are the most that people know of its interior. The building contains a sequence of splendid spaces culminating in the domed Grand Hall. The architect was E.W. Mountford and the building opened in 1907.

sympathetic reuse difficult. The best courtrooms are filled from wall-to-wall with eighteenth or nineteenth century fittings – benches, docks, witness stands, jury seating, judge's dais, public galleries. To remove these should be as unthinkable as the furnishings of the finest churches. In fact many of the problems associated with finding new uses for law courts are similar to those faced with redundant churches.

This report highlights the cases of nearly one hundred law courts from all over the UK. This is not in any way the full list of the great and the threatened buildings. Some of the most well known are not here – the Old Bailey and the Royal Courts of Justice for example; they offer fewer architectural revelations, and are under no threat. It does not consider some of the more unusual court types, such as Coroner's Courts, though certainly in London some fine examples are under threat, or the very rare and ancient surviving Manorial Courts, such as that which still convenes in a medieval room of Danby Castle on the North Yorkshire Moors. But it does include some of the very best, a selection of the most threatened and of the most representative. They serve as an illuminating introduction to, and celebration of, the fine but little known tradition of court architecture in this country and highlight the causes of redundancy and the possibilities and challenges of new use.

The report is based on three assumptions. First, that historic courthouses are a central part of our national heritage and an important part of the fabric of our towns and cities, built at public expense to serve the public. They should not be casually abandoned or destroyed.

Second, that they are best kept open for their intended purpose and that most are capable, with wit and imagination, of being adapted to meet current and future judicial needs.

Third, that when this is impossible to achieve, a suitable alternative use must be found, a use which respects the character of the building and the quality of the interiors.

The Scale of the Problem[*]

Over 800 law courts have been closed since the end of World War II. In England and Wales all types of court have been affected: Assize and Quarter Sessions, Magistrates' and County Courts. It is impossible to be certain how many purpose-built courthouses are amongst these, since Magistrates' Courts did not always convene in specialised buildings, and County Courts were often converted from buildings designed for other uses. But there can be no doubt that there are hundreds, and that perhaps the majority of the survivors will follow them into redundancy in the next few decades.

In 1971 assizes and quarter sessions took place in courthouses in 144 towns and cities. Of these buildings, only 36 are still in use by the Crown Court. Although some became Magistrates' Courts, such as Bedford and Appleby Shire Halls, most were made redundant. Those reprieved for the magistrates have not in the long term fared much better: Appleby has closed and Bedford is under threat.

The drop in the number of Magistrates' Courts has been even more dramatic. In 1945 there were 997 Petty Sessions Divisions (now Units, the basic unit of Magistrates' Court administration, consisting of one justices' 'bench'), and most of these would have had a building of some kind in which the court sat. By 1978 there were only 640 Magistrates' Courts and by 2002 this had fallen to under 400, many sitting in replacement buildings. This number is set to fall yet lower still as policies designed to improve efficiency and reduce costs are implemented across the country, a process of rationalization by which smaller courts are closed and facilities are concentrated in larger towns and larger court complexes.

The pruning of County Courts has been no less severe. The total stood at 493 on the eve of the First World War. By 1993 it was 274; by 2000 it had fallen further, to 230 (including a number operating in Magistrates' Courts). This figure will fall further too as a result of further rationalization and the widespread adoption of a range of IT initiatives.

[*] *Figures quoted in this section are reproduced from **The Law Court 1800-2000: Developments in form and function**, Allan Brodie, Gary Winter and Stephen Porter, English Heritage 2001*

Reform and modernisation: the causes of redundancy

The Crown Court: year round justice

THESE hundreds of closures have been driven by changes in policy, procedure and technology which have created demands on historic buildings for which they were never conceived. The first and most fundamental change was the abolition of the Assizes in 1971. Reform of the system of trial by jury had been discussed since the nineteenth century and by the 1960s the system of itinerant Assize judges, little altered since the middle ages, was under severe strain. Quite simply, the number of cases to be heard was far outstripping the capacity of periodic assize hearings to deal with them.

Appleby Shire Hall, 1778.
Home to the assizes until 1972, then the Magistrates until 2000. Currently empty, its future is uncertain.

The number of people on trial tripled in the thirty years from 1938. A Royal Commission headed by Lord Beeching (more infamous for closing railways than for closing historic law courts) recommended establishing Crown Courts that sat all year around (modelled on a system established in South Lancashire in 1956). These would be concentrated in the larger towns and cities, meaning the abandonment of many smaller county towns where the Assizes had sat for centuries. Some of these historic courthouses survived as Magistrates' Courts, such as Appleby and Huntingdon Town Hall, others, such as Presteigne Shire Hall, simply closed.

The fate of those in the towns and cities that were designated for Crown Courts was not much more certain. These were buildings which had been designed to handle far fewer cases, crucially for only a few weeks a year, and Beeching was very critical about conditions: severe overcrowding, little or no separation between the different users of the courts, the cold, the discomfort, the loos, the cramped and primitive holding cells. Naturally, the judge's accommodation was a subject of particular concern: 'the judge's retiring room may not be much bigger than a cupboard and may, indeed, serve the charwomen in that capacity when its distinguished occupant has gone'. The far greater facilities required for year-round sittings were simply non-existent in almost all cases.

In short, in 1971, no Assize Courts were in a fit state to serve

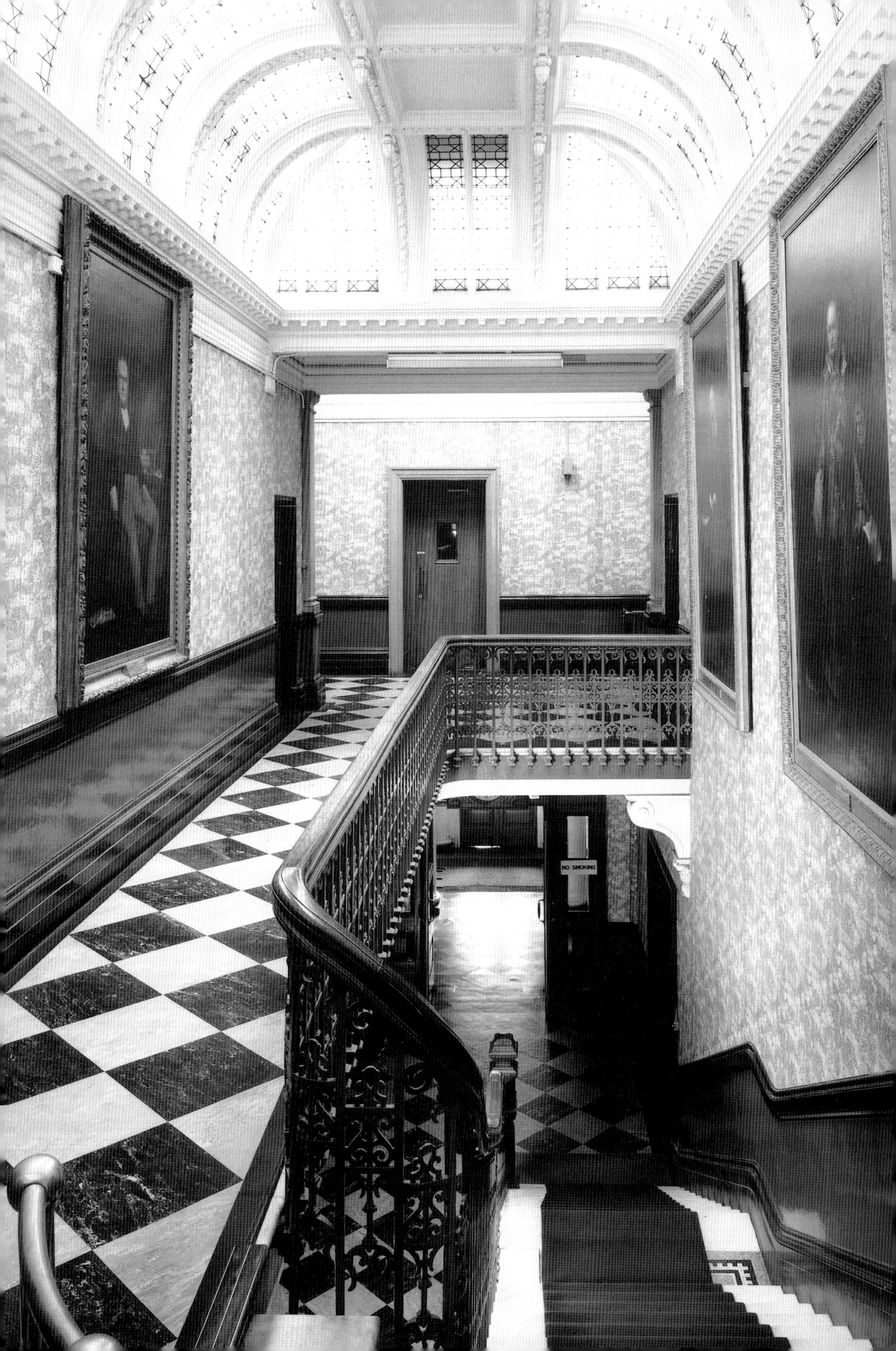
NO SMOKING

wes Crown Court.
ilt as the County Hall in 1802-12 by John
hnson, extensively altered in 1890-2, as seen
e, and commendably restored and
dernised by the LCD.

Manchester, Minshull Street.
The old and the new at Manchester's Minsull Street Crown Court. The Hurdrolland Partnership's 1996 scheme transformed a down-at-heel 1871 police court into a modern Crown Court complex. The key is the new glazed atrium.

as Crown Courts in the long-term. The response was the courts Building Programme, which ran from 1972 to 1996 and cost around £500m. 139 schemes were completed in that time, creating 382 new criminal courtrooms (a net increase of 212). It was central government's most prominent building programme of the time, commissioned by the Lord Chancellor's Department (LCD), which assumed responsibility from the local and county councils for the provision of Crown Court facilities in 1972, and was managed by the Property Services Agency (PSA).

The schemes completed under the programme were characterized by a number of concerns and concepts enshrined in a series of design manuals. First was, simply, the need to increase the number of courtrooms in order to cope with the increased court business generated by rising crime. Second was the requirement for strict segregation of the principal users inside the court: judges, defendants, jurors, vulnerable witnesses and the public and advocates. The need to keep these groups in separate areas with separate access to every courtroom created very complex planning requirements. Segregation was the key element of the increased security measures embedded in every aspect of these new courts. Another significant, and space-consuming, requirement was for extensive support accommodation – such things as the numerous consultation and meeting rooms and separate waiting and dining areas for the different groups. Information technology became an important issue too. Finally, a whole new type of courthouse emerged, the Combined Court Centre, in which facilities for both the Crown and County Court were combined in one building, bringing with it greater flexibility and efficiency. Twenty eight of these complexes, in which there may be as many as twenty or more courtrooms, were built.

Faced with these sorts of demands it is not hard to see why so many old and cramped Assize Courts were abandoned. Nevertheless, twenty two remain in use as Crown Courts today, proving that adaptation to meet these new requirements is possible. The challenges faced in attempting this, and the solutions that can be adopted to achieve it are discussed in more detail by Jim Stevenson in Chapter 4, but it is apparent from examples such as Lewis Crown Court and Worcester Shire Hall that with commitment and imagination even eighteenth century buildings can be sensitively refurbished.

A key determinant is the possibility of extending the building, since a lack of internal space is the principal shortcoming of most Assize Courts. Where there is room on the site to do so, such as at Worcester, Chester Shire Hall and Manchester Minshull Street (although this was built as police courts) creative solutions involving new wings and fully segregated plans have created viable Crown Courts. If there is little possibility of substantial enlargement, then the options are more limited. Warwick Shire Hall survives principally because the decision was taken to build a new Combined Court Centre directly opposite. Similar circumstances saved Newcastle's mighty Greek Revival Moot Hall – opened in 1812 – from closure. But Exeter Crown Court, occupying buildings in the castle where courts have sat for hundreds of years, is closing principally because there is insufficient room on the site to extend the building to the extent required. Some buildings, such as John Carr's York Crown Court or Thomas Harrison and J M Gandy's Lancaster Shire Hall, remain as lower level (so called Level Three) Crown Courts which require only modest improvements, possibly primarily for political reasons (Lancaster lobbied vigorously to prevent the removal of the court to Preston, and the closure of York would mean the nearest Crown Courts to the city would be in Hull, Leeds and Middlesbrough, not something the good burghers of York would easily accept).

For some of these court buildings, hanging on with only two or three courtrooms when the LCD really wants combined centres with half a dozen or more courts, the development of IT support for the courts system may prove their long-term saviour – by removing the administrative functions to regional centres and so freeing-up internal space for improved circulation, waiting and meeting arrangements they would be able to return solely to a role as trial centres.

Historic courts can have significant advantages over new buildings. Courtrooms and public spaces are normally lofty and airy, often top-lit, exactly the characteristics that the LCD now seeks in new court buildings after complaints about the low ceilings, artificial light and mechanical ventilation of many 1970s and 80s designs. Former Assize Courts also occupy prime central sites, with generally excellent public transport links. This is disproportionately important to many court users from deprived sections of society where 'transport poverty', specifically low rates of car ownership, is common.

There is also a powerful case for arguing that Government has a moral responsibility to make every effort to keep them in use as law courts. Assize Courts were built at public expense and maintained at public expense for public use. They have unusual and highly specialized plans which make sympathetic reuse difficult. Continued court use may be the most secure long-term future. Cynics may sniff, but this is a responsibility readily and seriously accepted by the Scottish Court Service.

Magistrates' Courts: the erosion of local justice

As the Court Building Programme wound down in the 1990s, the pressure to rationalize Magistrates' Courts was ramped up. Now all over the country local alliances of councillors, MPs, lawyers, magistrates and journalists are campaigning to save Magistrates' Courts from closure, fighting for what they argue passionately time and again is an essential local civic service, an important element of community life and the embodiment of the principle of local and visible justice.

A large number of closures since the Second World War have been sensible and logical attempts to iron out the peculiarities in the spread of Magistrates' Courts, the result of what often appears to be the irrational historical evolution of local justice. But there is no doubt that closures now are the result of cost-cutting and efficiency savings measures demanded by the Treasury, which in 1998 estimated there was 30% overcapacity in Magistrates' Courts across the country. Closures and amalgamations of justices' benches have also been accelerated by the decision to reduce the number of Magistrates' Court Committees from around 100 to 42 to bring them into line with the geographical regions of the police and the Crown Prosecution Service.

These pressures have been compounded by the increasingly demanding design requirements set by the LCD, particularly in terms of security, and custody cases have been withdrawn from a number of courthouses that no longer meet these more stringent standards. These concerns about security were triggered by the transfer of responsibility for prosecution in Magistrates' Courts from the police to the new CPS in 1986, which led to a substantial decrease in the police presence in courthouses.

The impact of this wave of closures has been felt most sharply in rural districts where the closure of courts in small towns has deprived many country communities of a local court for the first time in modern history. The LCD is at pains to point out that no Magistrates' Court is closed without careful analysis of the likely impact – statistics to prove that X % of the population live within Y miles of the new court; that Z % can reach it by public transport. But the plain fact is that the exercise is driven by the narrowest of 'value for money' criteria and little else. It is inefficient to staff a court for one morning a week, even if it is more convenient to court users. Key Performance Indicators are king – pounds per case and hours per courtroom per year. The latter is now at least 1,000. If MCCs don't meet these tough targets the harassed administrators managing them may find that their careers do not progress quite as quickly as they had expected. And these KPI targets only get tougher and tougher. One sore point is that the LCD measures the number of hours that the court actually sits, rather than those that were predicted and planned and provided for – many cases collapse and don't take up as much time as predicted, leaving courtrooms unavoidably empty. Staff have to be assigned and paid regardless, but the LCD does not take this into account.

Although decisions on the future of Magistrates' Courts are determined by the individual MCCs, not the LCD, the latter provides 80% of MCC funding and sets policy nationwide. As the Local Government Association has said, 'the LCD should not be allowed to shelter behind the decisions of the MCCs, because these decisions have to be taken within the financial limits and policy framework set by central government'. If an MCC wants to carry out anything other than minor improvements to a courthouse to bring it into line with current guidelines and inspectorate recommendations in order to keep it open, it must go to the LCD for the money. The power to grant or refuse this is one significant way in which the LCD can determine the actions of MCCs.

Nevertheless, to a large extent even the LCD's hands on funding are now tied. Its direct grants to the MCCs today are only available for relatively small-scale refurbishment. Major capital investment has to be delivered by the Private Finance Initiative (PFI), but this inevitably leads to the concentration of courts in fewer buildings because the PFI system, as discussed below, is structured against the retention of small Magistrates' Courts.

And yet, particularly in rural areas, such rationalization directly challenges the LCD's proclaimed intent of improving the quality of justice, which as it is often keen to emphasize, means improving people's access to justice. This means that rationalization is counter to the government's social inclusion agenda. Increasing the distance that people have to travel to court disadvantages those sections of society most dependent on public transport – those from the most deprived sections of society who suffer most from transport poverty – and yet these are exactly the people that are most likely to go before the courts. Those most in need suffer most.

This problem is particularly acute in rural areas where public transport is often bad or non-existent. An MCC may claim in justification of a closure that, for example, travel by public transport to the alternative venue takes less than an hour, but the bus timetable may only have a frequency of two hours or worse. What is the logic in closing a court for amongst other reasons inadequate disabled access only for disabled users to have to travel many extra miles to get to the replacement? And greater distances make getting witnesses and defendants to attend more difficult, leading to listing difficulties and more delays, increasing case costs and time.

Closure of rural courts runs counter to Government policy for rural areas. Its recently published blueprint for the future of the countryside recognized that 'market towns… must be the focus of a range of private and public services to which people need access'. Amongst the facilities which towns with a population of between 10,000 and 25,000 require under the blueprint is a Magistrates' Court. Yet more and more such towns are losing theirs.

Moreover, an MCC may save central Government money by closing courthouses (because the centre grant aids the vast majority of an MCC's costs), but these savings to the public purse are far outweighed by the extra strain in time and cash resources borne by the police, probation services and other public officials faced with longer journey times to the replacement courthouse. This is not the only false economy: magistrates, witnesses, defendants and lawyers are also burdened by the extra travel time and expense.

And yet these consequential costs of closure are not taken into account in any meaningful way. The closure report published when the court in Haverhill in Suffolk was shut down recently identified potential annual savings of £11,160, but acknowledged that closure would mean additional costs 'may be placed upon other organisations and individuals' without actually taking these into account.

In addition there is the impact of closure on the local economy: business serving the court losing valuable custom, and legal practices leaving town in the wake of the court. And at a fundamental level, this rationalization challenges a basic principle behind the English justice system - that (as far as possible) justice should be locally dispensed by local magistrates.

It is not surprising, then, that County and District Councils and local communities fight closure vigorously and frequently. Councils are allowed to appeal formally against proposed closures, but they face a formidable and intractable opponent. Cumbria County Council lost its campaign to prevent the closure of four courts in 2000, but in 2003 sustained opposition to the planned closure of Bolton and Salford Magistrates' Courts was finally successful. Over 21,000 people in Bolton signed a petition organised by the local paper against a proposal that, extraordinarily, would have left a borough with a population of 260,000 without a Magistrates' Court.

County Courts: a coming IT revolution?

Although there have already been hundreds of County Court closures there are likely to be many more to come. The process may even accelerate. To date it has been a matter of 'rationalizing' courts in smaller towns and moving into Combined Court Centres in larger urban areas. However, the Court Service* consultation paper, *Modernising the Civil Courts* (2001), provides a glimpse of a future for the civil courts in which the possibilities of information technology and the internet greatly reduce the need for physical court facilities. The nature of civic court business means that the potential for these new technologies to bring about radical change in the way court business is conducted is much greater in County Courts than in the criminal justice system.

The baseline development is the digitization of all court records, reducing the need for both administrative accommodation and staff and allowing them to be concentrated more efficiently and at less cost in regional support centres rather than in every individual court building. But there are a whole raft of other possible developments beyond this, many of which are being tested at the moment: online

---

* *The Court Service is the executive agency, an arm of the LCD, established in 1995 to manage the Crown and County Courts and the LCD's substantial responsibilities for Magistrates' Courts.*

Cheltenham County Court.
Thomas Sorby's 1871 building is a cut above the rest. It is also the best preserved.

applications to courts and issue of claims, judgements and warrants; video conferencing; electronic presentation of evidence; 24 hour information services via call centres and internet sites. Already in some US states you can get a divorce over the internet.

What appeals to the LCD about the widespread adoption of these technologies is the prospect of creating a 'virtual court' which would allow the LCD to take the court to litigants rather than forcing on them the inconvenience of physically attending the court, something of particular relevance to rural communities. This is about improving customer service. Probably as importantly, however, it could result in huge efficiency savings for the Court Service, because if these ideas work there will be a substantial reduction in the number of courtrooms and local office space required (though an increase in centralized facilities).

At the same time the Court Service is undertaking a thorough review of the County Court network to identify substandard facilities and areas of under and oversupply. The aim is to 'restructure the civil court network, providing courts according to regional need, population distribution and transport networks rather than historical accident'. Combine these two objectives and what you get, of course, are widespread closures of County Courts.

This in itself should not present an overwhelming conservation challenge. Many County Courts are housed in converted buildings, typically post-war office blocks. There are no purpose-built County Courts older than 1846 (when the system was created) and those built thereafter, though often handsome buildings, are rarely outstanding civic statements like so many Assize Courts. This reflects their lower status in the legal system (cases could be referred to the Assizes as the superior court). Their planning is less sophisticated and specialist than Sessions and Assize courthouses: there is no need for cells or secure access to courtrooms and complicated circulation to separate parties; courtrooms themselves are more simply planned, with simpler furnishings and only limited staging and changes of floor levels. The County Court is much more like an office building, making sympathetic adaptation a simpler proposition.

The Private Finance Initiative: the future of law courts

As was touched on above, the Private Finance Initiative (PFI) is now almost the only means of providing improved court facilities in most of the UK. However, the very nature and structure of PFI is such that the process itself, as much as policy, is determining the extent to which historic court buildings are now being closed.

PFI is the highly contentious form of Public-Private Partnership (PPP) by which Government contracts out the provision and modernization of public services to the private sector for a fixed term (normally twenty five years) in return for a fixed annual fee. It has been endlessly discussed, debated and analysed, but the widespread criticisms, now well worn, are worth repeating: that the driving force is, not is as claimed, to introduce private sector efficiencies to the provision of public sector services, but to remove the costs of modernizing public services from the public sector borrowing requirement; that any savings that might accrue from the greater efficiencies of private sector practices and management are cancelled out by the higher costs of private borrowing against the rates the government can borrow at, by the profits expected by the consortia, the sub-contractors and all their shareholders, and by the millions that are spent by consortia bidding for every contract – all of which are transferred to the public in the size of the annual charges; that as a result there is a mounting body of evidence that PFI is often more expensive than traditional direct expenditure; that in reality there is no transfer of risk to the private sector, because if it comes to it the government would have to step in to ensure public services were maintained; that the public sector comparator, the test by which government assesses whether a PFI scheme represents better value than traditional direct procurement is an infinitely flexible and therefore meaningless exercise which can produce any answer the Government might wish (as was shown in the London Underground PFI schemes); and that the idea of public buildings as civic projects standing for something greater than the most basic functional necessity has been replaced by a profit-driven ethos that strangles aesthetic considerations and even functionality, creating buildings that are at best uninspiring and at worst functionally flawed.

The charming Magistrates' Court in **Keswick** was one of four closed in Cumbria in 2000 despite widespread protest.

In June 1996 the Government announced that henceforth all

new courts would be built using PFI. So far the impact is being felt most in Magistrates' Courts, which are the current focus of the LCD's modernizing and rationalizing agenda. Here the very structure of PFI is having a dramatic impact. The large consortia who dominate PFI have no interest in small-sized contracts to run one or even a few small Magistrates' Courts, and certainly not ones that sit empty for most of the week. These simply do not offer attractive returns, not being valuable enough to justify the extraordinary expense and effort that the PFI bidding process involves. The consortia seek projects involving large court centres or bundles of multi-courtroom Magistrates' Courts, and this, not the MCCs, or even the LCD, is what is shaping Magistrates' Court PFI projects. To make them large enough to be commercially attractive they are organized to include all the facilities in one MCC area, or even two small ones combined. This leads to the closure of small rural courts that otherwise might have been kept open by MCCs through relatively small-scale incremental investment in modernization. PFI ties clients such as MCCs into predicting accommodation demand for the next twenty five years, and for paying for it in twenty five years time whether it is still all required or not, whereas alternative systems – such as the previous regime of direct grants from central Government – would allow MCCs to refurbish and build courthouses in a manner that matched much more closely actual needs as they change and evolve over time. The net result, for example, under the Derbyshire MCC PFI scheme (one of the first) is that the number of towns in which courts will sit will be reduced from twelve to four.

PFI also brings into question the long-term future of smaller historic Crown and Combined Courts, buildings like York or Lancaster. The private sector's distaste for smaller single-facility contracts would surely include the refurbishment of these buildings, particularly because they bring with them restrictions imposed by listing. Instead, future modernization would have to be carried out in-house, through direct expenditure. But government departments do not have much freedom to operate in this way anymore. The recent reprieve for the Crown Court in Knutsford Sessions House – another victory for determined local campaigning – which was to have closed in the face of a multi-million pound repair bill and the need for complete modernization, suggests that money can still be found for major capital projects in the specific and restrictive circumstances that pertain to most of the architecturally important Crown Courts still operating.

However, the LCD and the Court Service have at least the advantage over many PFI clients, such as local education authorities, of being a large centralised body able to bring substantial amounts of the skills, savvy and experience to the incredibly complex PFI process necessary to derive the maximum benefit from it. Although individual MCCs are in fact the clients for Magistrates' Court schemes, the Court Service provides valuable support and advice and in two important areas it is trying hard to overcome some of the limitations of the process. The first concerns listed buildings. Central to private sector service providers maximising their profits is their ability to keep the costs of operation and maintenance over the life of the contract to the lowest possible levels. With a fixed price contract, the last thing they want is unexpected and expensive problems to emerge in a building in twenty years time. At the same time (as the process was conceived) consortia are free to provide the services contracted in any manner they choose, providing it meets the terms and conditions of the contract. This inevitably means that in order to control building costs they prefer to build new buildings than modernize older ones, especially listed buildings which they rightly or wrongly perceive to have high and unpredictable refurbishment and running costs. PFI therefore threatens to bring a rapid end to the policy of modernizing Georgian and Victorian Courts which at Manchester, Chester and other locations had been such an admirable success.

Derby Shire Hall, a Grade I listed building with a seventeenth century core, has been the first major testing ground. When it became clear that the consortia bidding to provide a new Magistrates' Court for the town had no intention of refurbishing the Shire Hall if given the choice, the PFI bidding process was halted by the Court Service, which introduced reconsideration of the Shire Hall as a mandatory variant of the contract, telling bidders that the LCD would increase grants to bridge the extra cost of this option, although, if the site could not physically provide the facilities required by PFI, they would not be forced to include it in their bids. The scheme had become a

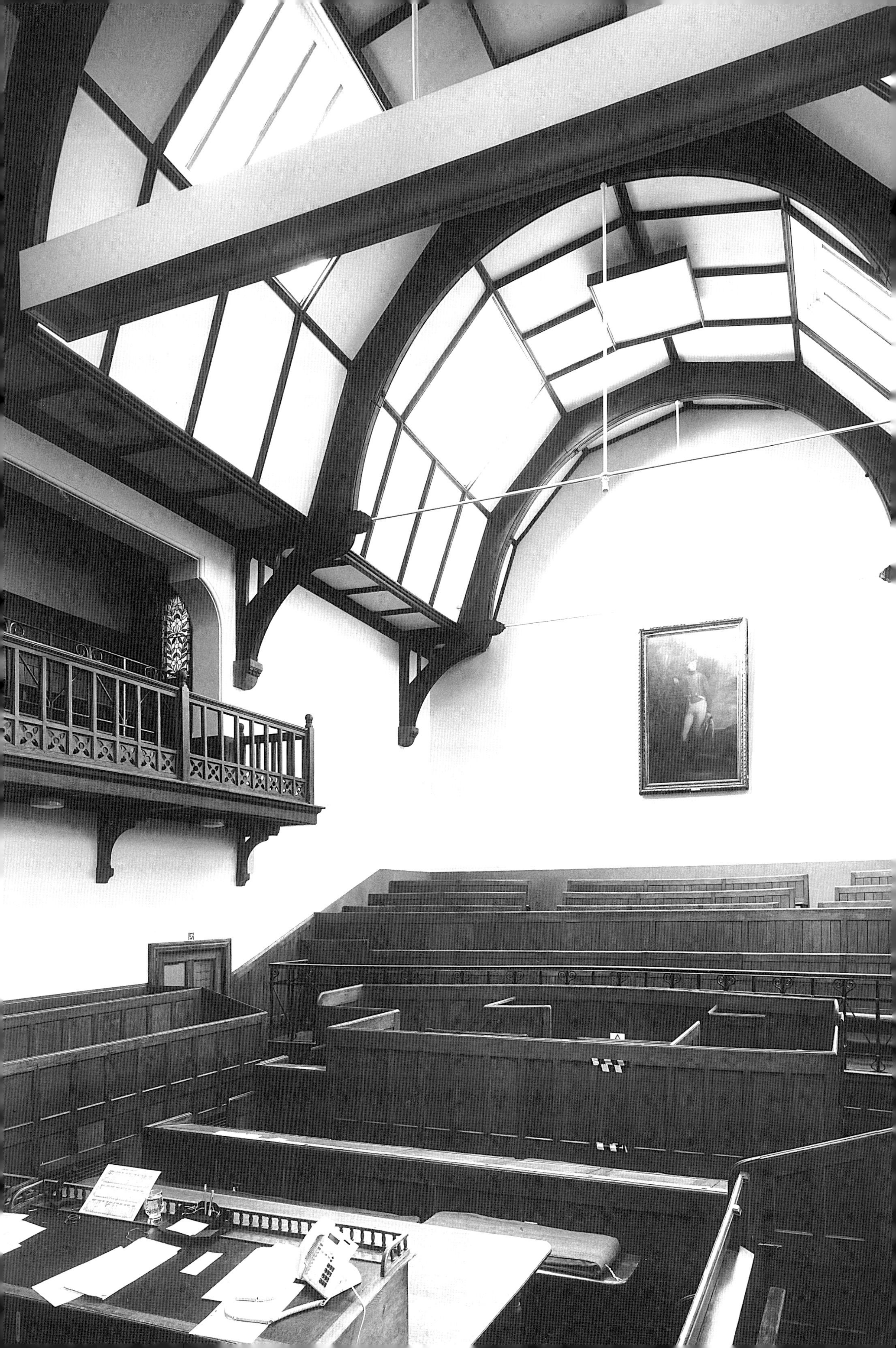

red Waterhouse's **Bedford Shire Hall**, w a Magistrates' Court, but will it vive the PFI process?

'pathfinder project' which would be used to identify means of marrying the particular requirements and responsibilities inherent in government ownership of listed buildings to the PFI process. The changed rules succeeded in forcing the bidders to include the Shire Hall in their proposals, and the winning consortium has almost completed work on a scheme that is creating a further ten courtrooms at the site – a scheme which, although sadly architecturally woeful, will at least bring the decaying building back into use.

Derby is important because the LCD and Court Service need to be able to demonstrate through it to bidders for future projects that bringing together complex listed buildings and PFI is viable. MCCs, as the clients, need to be persuaded too. They may equally view run down historic courts as the problem, not the solution. Faced with a Victorian building that has been starved of investment for decades, it is unrealistic to expect them to fight for its refurbishment against the shiny new complex proposed by the bidding consortia. The future of important buildings affected by PFI schemes, notably Waterhouse's Bedford Shire Hall and the magnificent Victoria Courts in Birmingham, may be dependent on the success of the Derby experiment. However, the Court Service has had to expend a vast amount of effort, and secure special additional funding, to ensure that Derby is refurbished under PFI. It is difficult to imagine that it will be prepared, or able, to do the same for less outstanding listed buildings. Casualties should be expected. The LCD has also gained something of a reputation for trailblazing in its attempts to address one of the most stinging criticism of PFI, that it delivers mediocre buildings and positively inhibits good design. Many observers still believe that despite some fine words a mentality prevails both at the Treasury and bidding consortia that sees 'design' as an almost irrelevant add on, not integral to creating efficient and successful buildings.

Such an attitude certainly does not exist at the top of the Court Service. It has reshaped its PFI guidelines to give more weighting to design in the selection of contractors. It also plans to pay 'honorariums' to the architects that finish second and third major in PFI bids in answer to complaints that smaller architectural practices are being squeezed out of the PFI market because they cannot afford to take the risk of committing months and even years to a project, working on discounted fees, only ultimately to lose. It has issued a statement on design quality for bidders, which says all the right things, and collaborated with the Commission for Architecture and Built Environment to pioneer the idea of a department 'design champion' (in this case Ian Ritchie), a concept which is being adopted across Whitehall as part of the 'Better Public Buildings' initiative to improve the quality of public sector buildings.

Ritchie's brief is to assist the LCD in ensuring new buildings meet higher design standards as part of a new two stage PFI process. Under this, once the LCD has drawn up a shortlist of bidders – who at this stage have not appointed architects – they are given presentations from Ritchie on the importance of good design, and from the LCD on the history of court design and operational requirements, and from the local authority on its own aspirations for the project. The consortia then team up with architects to produce design concepts before the final three are selected. Only then do they work up fully-costed schemes, based on the earlier concepts, before the successful bidder is chosen.

Five projects have been chosen to pilot the reforms including the Bristol and Manchester Civic Justice Centres and the East Anglia Crown Court scheme, for new courthouses in Cambridge, Norwich and Ipswich. Privately, the Court Service admits that it has set itself a very difficult task. Although the political will exists at the highest level – the last Lord Chancellor of course was famous for the deep interest in design he showed at his own Palace of Westminster offices – the hardest constituency to convince is the senior administrators, trying to balance extremely tight budgets and manage the process as clients. They have yet to be convinced that this new emphasis on design brings benefits that can justify extra financial cost.

The most important of the pilots is the Manchester scheme; with forty courtrooms and hearing rooms it will be the largest courthouse to be built in the UK since the Royal Courts of Justice over a hundred years ago. For this the LCD ran an architectural contest independently of developers to prevent the design from being subjugated to other financial considerations during the bidding process, another pioneering PPP arrangement. The successful practice was the Australian firm, Denton Corker

Newcastle Combined Court Centre, 1990.
By the Napper Collerton Partnership.
The LCD was very proud of this product of the Court Building Programme, which occupies an important site on the Quayside close to Wilkinson Eye's winking bridge – but it is a ponderous, unsubtle brute.

Marshall. Whether or not the LCD will be able to force the developer Allied London to implement the winning design without it being significantly diluted is another matter. Let's hope so.

Many people are watching this project closely to see whether this strategy will produce a building to match the quality of the Royal Courts of Justice. The standard of courts built in the last thirty years, particularly considering that they are the most visible group of buildings erected by central government in that time, has been sadly and generally dispiriting and disappointing. Successes such as Truro Combined Court Centre and the recently completed (but pre-PFI conceived and funded) Southampton Magistrates' Court by Hampshire County Council Architecture and Design Services stand out because they are the exception, not the rule. And the designs produced for some early law court PFI projects have done little to suggest that the process will do much to raise standards: Hurdrolland Partnership's recently opened court centre in Belfast has ambition and impressive public spaces, but many schemes are bland and nondescript, and that for Derby's Grade I listed Shire Hall, a very sensitive project, is crude and heavy-handed. It is significant that, whereas many of the finest Georgian and Victorian courthouses were designed by the leading architects of their day, men like Street, Soane, Waterhouse, Harvey Londsdale Elmes, Smirke, Adam, Carr, Dobson and Thomas Harrison, no courthouse of the last thirty years has been designed by one of our best architects. This is despite the fact that two current British architects of international standing have designed law courts in Europe in recent years: Richard Rogers has completed courts in Strasbourg and, to acclaim, in Bordeaux; David Chipperfield is designing courts in Sicily and Barcelona, both of which look on paper to be superior to any courthouse erected in the UK in the last few decades. Many of the most prestigious architectural practices are reluctant to get involved in PFI, possibly because there have been enough (Lottery funded) public commissions around in recent years not to have to get mixed-up in a complex, involving process in which their ability to assert control over design is far from certain. But we need them to.

However, with the construction of the Manchester Civil Justice

Centre eagerly awaited and Fielden Clegg Bradley selected as lead architect for the Bristol scheme, the current round of PFI projects might mark a turning point in the quality of new courthouse architecture. The LCD and the Court Service certainly deserve praise for their efforts to overcome the dangers of PFI.

Law courts are major public buildings. That places on their design demands and expectations well beyond the average speculative office development. Not only should they be built to function immaculately and clearly and be a pleasure for staff and users, but more than that they should to be built with a generosity and dignity that expresses our fundamental commitment as a civil society to the rule of law. As civil buildings they have an obligation to enhance public life and public space. Nothing less is acceptable if the 'Better Public Buildings' initiative is to be anything more than hot air.

Planned law court complex, Barcelona.
David Chipperfield is one of Britain's few architects of truly world standing and yet neither he nor the likes of Foster, Rogers, Grimshaw, Hadid or Alsop has ever built a courthouse here. Chipperfield is building two on the continent though, including this ambitious scheme.

Northern Ireland: politicised buildings

COURTHOUSES in Ireland share many similarities in terms of planning and stylistic development with their counterparts in England and Wales. This, of course, is no coincidence: under colonial rule from London an English legal system was imposed centuries ago. It has resulted in some very fine buildings, none more so than James Gandon's Four Courts in Dublin, the most magnificent classical courthouse in the British Isles.

Today in Northern Ireland the judicial system remains closely tied to England and Wales, under the authority of the Lord Chancellor's Department in London. Administration is carried out locally by the Northern Ireland Court Service (NICS) in Belfast. At the head of the court structure is the Supreme Court of Judicature of Northern Ireland, made up of the Court of Appeal, the High Court and the Crown Court. Below that there are County Courts and Magistrates' Courts. In all, there are twenty five law courts operating or under construction in the province.

The Court of Appeal and the High Court sit in the Royal Courts of Justice, an impressive building of 1933 by J G West, in that heavy pre-war Imperial idiom that is inevitably, as here, finished in Portland ashlar. The Crown Court, County Court and Magistrates' Courts often share courthouses, typically with four courtrooms – effectively mini Combined Court Centres. Because these each serve much smaller populations than their English equivalents and thus require fewer facilities, it has been possible in many cases to keep them in nineteenth century assize and sessions buildings, even though these were normally constructed on a smaller scale than most English examples. The repair and refurbishment of Londonderry and Armagh by the NICS are recent successful examples of this.

Nevertheless, the inexorable process of rationalization has made many courthouses redundant over the last few decades, particularly the smaller nineteenth century Petty Sessions' Courts. A number of these are still at risk, but there are also some successful examples of adaptive reuse, such as Hillsbrough and Markethill, which have both recently been restored as lively community centres with the help of Lottery and other public funding. These are excellent Lottery schemes, transforming prominent derelict eyesores into centres of those in village life.

A number of larger courthouses have also been made redundant, replaced by new facilities. Antrim, dating from 1726, is the oldest still standing in the Province. Since a new courthouse was built it has been well cared for by the local council. Lisburn's impressive 1884 courthouse was demolished unnecessarily in 1971. Coleraine, a Doric prostyle temple of 1852, was closed in 1988, and stood empty for twelve years until J D Wetherspoon opened it as 'The Old Courthouse' pub in 2001. Easily the most important though is Crumlin Road, Belfast, the grand nineteenth century Crown Court begun by Charles Lanyon. It has been been replaced, under a PFI contract, by a new Combine Court Centre at Laganside and its future is now the responsibility of the PFI provider, Dunloe Ewart. Current proposals would see the retention of the front two bays of the building and one of the court rooms with some sort of display, and a large commercial office development attached to the rear. There must be a more appropriate solution than this for the building.

To this list of closures the NICS's *Accommodation Strategy 2001-10* threatens to add more buildings. Under this blueprint the NICS is closing courts identified as having inadequate facilities that cannot be easily improved. Equally important, no doubt, is the desire to save money. This, and the fact that new and refurbished facilities will be provided by PPP/PFI means that smaller courthouses are being closed (including all the remaining single courtroom buildings) and sittings concentrated in a smaller number of larger buildings. In total eight are slated for closure, an exercise that the NICS estimates will save £300,000 in running costs and capital charges each year. Once again local justice is being eroded, essentially, to save money.

Armagh Courthouse after the devastating 1993 car bomb. It has since been rebuilt, and reopened in 1999.

Not all these buildings are attractive or significant, but in particular the Italianate Banbridge (1874), the Germanic Limavady (1914) should be sympathetically reused.

These issues and these processes are similar to those that affect historic law courts in England and Wales too. They are not, in that sense, exceptional. But in Northern Ireland there is one extra and significant dimension that has had a dramatic impact on courthouses, and that is, of course, the Troubles. Courthouses throughout Ireland occupy a far more resonant, more highly-charged, position in the fabric of towns and cities than their English or Scottish counterparts because they were built as expressions not only of the Crown's power and authority, but of imperial rule in a conquered land. Thus, from the beginning of the Troubles in Ulster, as an embodiment of British rule and Protestant ascendancy, they have been a part of the struggle itself. The prominent role courthouses have played as the stage for the trial and conviction of terrorist suspects has only made their political symbolism more potent. As a result they themselves have frequently been the target of terrorist attack and a number have been very heavily damaged by bombs.

The architectural response to this has varied. Londonderry and Armagh have been restored after bomb attacks. At Armagh the whole of the interior was destroyed, but the decision was taken to recreate it, including decorative plasterwork and oak furnishings in the courtrooms. To some in the Loyalist community this was essential: anything else would be seen as a surrender to Republican terrorists. Dungannon, on the other hand, was abandoned after it was bombed in 1989. Crumlin Road is loaded with particular political potency, as Belfast's Crown Court and scene of the notorious Diplock trials. Though the condition and limitations of the building might have proved insurmountable barriers to retention and restoration as part of the Belfast PFI scheme in their own right, there must be no doubt either that the building was so

21

deeply scarred by divisive political symbolism that the NICS and LCD preferred to start afresh in a new complex unburdened by such a potent and controversial past. The same issues may also hamper efforts to find an appropriate new use.

The bombings and terrorist trials have placed security demands on historic court buildings far more stringent and challenging than faced elsewhere in the UK. At Armagh and Londonderry these have been met remarkably sensitively, though inevitably they have required compromises that would be unacceptable in other circumstances. For example, the appearance of Georgian glazing at Armagh has been kept up by gluing false glazing bars to the 35mm thick bomb-proof glass. The architects felt this was a lesser evil than installing secondary glazing, which would upset internal proportions. Particularly successful has been the security fencing in front of both of these courthouses which meets NICS requirements without shutting off the buildings from street. They remain, as they must be, an important element of the public domain. The dreadful alternative is Newry, where the Courthouse has been brutally segregated from the town by a twenty foot wall. With only the top of the building visible above it, this wall makes the decision to retain and restore the 1841 façade, when the building was otherwise completely rebuilt in the 1990s, seem slightly perverse.

The potent symbolism of Ulster courthouses remains a sensitive political issue in the era of the Peace Process and stuttering devolved government. Legislation soon to be introduced will require the removal of the Royal Coat of Arms from courtrooms because of the offence they cause many in the Nationalist community. The move is intended as a statement of the impartiality of the legal system and is described as a symbolic break with the past. It has predictably angered Unionists. The Ulster Architectural Heritage Society has asked for it to be waived in the case of historic or high quality examples, which will instead be screened by curtains. Royal arms will remain on building façades.

Scotland: a different path

SCOTLAND has a long and proudly independent legal tradition, something which is reflected in both in the architecture of its court buildings, and in its approach towards their continued use.

The country has a three-tiered court system. At the lowest tier are the District Courts, run by local authorities and rarely occupying buildings of any architectural interest. These equate to Magistrates' Courts – summary courts with a bench of local magistrates. Next come the Sheriff Courts, administered by the Scottish Court Service (SCS). There are forty nine of these throughout Scotland, the vast majority substantial nineteenth century buildings which took over the court function previously exercised by the 'tolbooth', that distinctively and uniquely Scottish civic building type with its prominent steeple that was the centre of local administration and justice from the middle ages. Alongside the earlier tolbooth, and possibly combined with a town hall, the Sheriff Court is often the most important architectural expression of civic purpose and ambition in a Scottish town or city. They function like Combined Court Centres in England and Wales, dealing with both civic and criminal cases tried by jury, though because of the smaller populations they serve they are normally smaller.

Above them are the Supreme Courts of Scotland – the High Court dealing with the most serious criminal cases and the Court of Session with the equivalent civil cases. These courts are also administered by the SCS and sit at Parliament House in Edinburgh, the fascinating complex which has been the home of the highest courts since the seventeenth century. The High Court also sits in the Justiciary Buildings in Glasgow.

Architecturally Sheriff Courts reflect the wider stylistic trends of the country. The surviving

Forfar Sheriff Court, 1871.
Typical Scottish court joinery.

early nineteenth century buildings, such as Perth, are gravely Neoclassical; later in the century, Scottish Renaissance and Baronial styles become common (Stirling for example). Whilst in some respects the planning of Sheriff Courts shares similarities with contemporary English courts, there are frequent differences. There is often only one courtroom, originally dealing with both criminal and civil cases (such as at Stirling and Dundee), unlike the separate courtrooms standard in English Assize Courts. This, presumably, reflected a lower level of court business. There is often a marked difference in the layout of the courtroom too. Some, such as at Perth, or the Justiciary Buildings in Glasgow, are immediately familiar to English eyes, with public galleries supported on iron columns, a dense layout of compartments laid out around the well of the court and the judge's bench close up. They are intimate auditoria. But there are also many courtrooms, such as Forfar's, that have a much calmer and less adversarial atmosphere. The familiar elements of the plan – the bench, table, dock, witness stand, jury box, benches for advocates, press and public seating – are all there, but the rooms are more spacious, nearly always rectangular, and light and lofty. There is far less variation in level: the bench is raised, but not always greatly; so too the jury box and witness stand, but only slightly. The public seating may have a shallow rake or none at all. Everything else is at floor level. The furnishings can be more generously spaced and the dock without the high sides, railings and screens deemed essential for security in England.

About 80% of the SCS's estate is listed. That means that the vast majority of historic Sheriff Courts buildings are still operating as courts, in contrast to England and Wales. The SCS suggests that this is primarily because Scotland has witnessed far less dramatic changes in population distribution and size. Thus old court buildings are still located where they are most needed. In addition the greater distances between the populations served by Sheriff Courts reduces the opportunities for rationalising facilities without imposing unreasonable inconvenience and expense in terms of time and travel. The courts themselves, again because of the smaller size of the population, have not been faced with as dramatic an increase in court business as many of their English counterparts, and the attendant pressure for additional accommodation. Nevertheless, they still face demands common to other court buildings in the UK: the need to increase the number of courtrooms, upgrade public areas, improve security and custody suites, fully segregate circulation, install IT networks, and increase the number of consultation rooms. The SCS is coming towards the end of a comprehensive programme of works to enhance facilities at all courts, and yet on only one occasion has it done so by closing the existing courthouse – at Kilmarnock, where the listed 1850s courthouse was converted into the office of the Procurator Fiscal. Everywhere else the SCS has chosen to restore and refurbish existing listed courthouses, extending them to create additional courtrooms and accommodation if necessary.

The result has been a series of generally excellent schemes, such as the Justiciary Buildings, Glasgow and Dundee, Paisley, Dumfries and most recently Ayr Sheriff Court, frequently carried out with the confidence to avoid pastiche or historicism. The SCS says that it prefers refurbishment to replacement because the existing buildings are usually very centrally located, convenient for public transport and professional court users and also (and significantly) because it has a 'moral obligation' through the Scottish Parliament to maintain in good repair listed buildings in its ownership – and the best way to do that is to retain them in their original use. It is refreshing to hear this so unambiguously stated by a government agency, and actually adhered to in practice.

Undoubtedly this is often easier for the SCS. Courts such as Dumfries require less extensive expansion than is the norm in England, which makes refurbishment of listed buildings on constricted town centre sites far less problematic and expensive. Nevertheless, there is no doubt that the SCS is committed to the policy because it is proven to deliver high quality facilities that are appreciated by users and improve the effectiveness of the judicial system. Through experience it understands that first class courthouses can be created in historic buildings by working sympathetically with them.

Interestingly, the SCS has yet to dabble with PPP. Its refurbishment projects are efficiently run and effective, and anyway single courthouse projects on the scale of a Sheriff Court, unless supported by commercial development, are simply not commercially attractive to the private sector. The only PFI-sized project so far, the £105m refurbishment of Parliament House, is being managed in-house.

Dundee Sherriff Court.
Nicol Russell Studios refurbished and extended the 1863 building in 1996. Extra accommodation was created by subtle, contemporary intervention.

Redundancy: preservation or reuse?

WHAT happens to the hundreds of courthouses made redundant by rationalization and policy changes? Many have found new uses, some sympathetically, some insensitively. Others are empty and decaying. A number of the most historic have been opened to the public as museums.

Morpeth Sessions House, 1828.
John Dobson's splendid Sessions House is now home to a women-only fitness centre; tanning booths occupy the bench!

In fact, the number of redundant courtrooms which are preserved and opened as exhibits comes as quite a surprise. This may partly be explained by the fact that courthouses are civic buildings, owned by public authorities, to which a large amount of civic pride may be attached. It may also be because, if the costs of maintaining the fabric are set aside - if for example the courtroom is a part of a town hall and the owners are therefore committed to the long-term ownership and maintenance of the building anyway – the cost of running such exhibitions is relatively low. Most merely consist of a few cases, explanatory boards and mannequins or cut-out figures.

However, the Galleries of Justice in Nottingham has shown that with more investment such exhibitions have enormous potential well beyond simply illustrating how justice was dispensed centuries ago. The key here is the interactive mock trial. This has proven to be a big hit with revenue-generating corporate clients. It can be used as part of team development work or as entertainment: few workforces like anything better than the chance to put their bosses on trial. More significant is the mock trial's immense popularity with schools (50% of visitors are school parties), as part of history classes or citizenship lessons (or anything else for that matter – you can, and schools have, put any issue on trial). The Galleries are developing re-enactment packs for use by museum education officers at other preserved courthouses. They have also taken the potential of the museum and developed it into a National Centre for Citizenship and the Law, which has expanded the education programme by focussing on how a cross-curricular approach to issues of citizenship and personal responsibility can be addressed with interactive trials, discussion and video recording and editing of proceedings. The Centre works in particular with children who have been excluded from school and young offenders, attempting to get them to look at their own behaviour and its impact on others, and to understand what their rights and responsibilities are, through film, debate and workshops. Funding has come from the Home Office and the Treasury amongst others, and the initial evidence suggests that the majority of those who have attended have not re-offended. This remarkable scheme has recently been recognised with the ward of the £100,000 Gulbenkian Prize.

Clearly not every preserved courthouse can be host to such a centre, but there is no doubt that the concept could be exported around the country. Many conserved law courts – places such as Stafford, Northampton, St George's Hall Liverpool and

Maldon Moot Hall.
A fascinating building that dates back to the fifteenth century and contains a charming little Georgian courtroom.

Carlisle – with advice and support from Nottingham could offer similar citizenship programmes to schools, and host income-generating corporate mock trials. Nottingham has demonstrated that these activities are not only compatible with conserving a listed courthouse interior, and wider public access, but can also generate valuable revenue to support ongoing maintenance and repair, something of vital importance to cash-strapped local authority owners.

But the preservation of law courts, intact, necessarily only occurs in a minority of cases. Most redundant courthouses are adapted to new uses. Perhaps the single most important determinant of fate is type – what court was the building designed to house, and is it a single purpose courthouse or does it form part of a multipurpose building such as a town hall or police station? This will dictate its suitability for adaptive reuse. Thus rural Magistrates' Courts, often purpose-built as Petty Sessions Courts, can relatively easily be converted to new uses, frequently offices. They rarely have extensive or high quality furnishings, the planning is simple, the courtroom modestly-sized.

This is also true of Victorian and Edwardian Police Courts attached to police stations. If the police station is being closed as well, a common occurrence, then the whole building may become available (for example, Chipping Campden, Gloucestershire). If it isn't, this might impose restrictions in terms of parking, curtilage or security, though there are a number of examples to show that this does not have to impede reuse (such as Lutterworth in Nottinghamshire). Often, because these buildings are owned by the County Council or Constabulary the building is recycled as a publicly owned asset, becoming, for example, a probation service office, or a 'one stop shop' for council services (Keswick, Cumbria and Thornbury, Gloucestershire), or the registry office (Wigton,

Cheltenham Crown Court.
With an all but perfectly preserved interior, this courthouse is a strong candidate for conservation.

Warrington County Court.
Henry Tanner's courthouse has recently been imaginatively converted into an arts centre, with lottery funding.

Cumbria). Where, commonly, Magistrates' Courts sat in town halls or guildhalls, often in a room that doubled as a council chamber or assembly room, the withdrawal of the court has rarely had a significant impact on the future of the building. Older examples are often recognized as significant local historic buildings, cherished expressions of the history and heritage of a community, and proudly preserved by town councils. The Moot Hall at Maldon, Essex, is a wonderful fifteenth century brick tower with a Georgian porch and charming Lilliputian interiors, including an eighteenth century courtroom with panelled dock, bench, witness and jury boxes, all perfectly preserved since the magistrates left. The ground floor courtroom of Sandwich Guildhall is also lovingly preserved, along with fittings that may date from its construction in 1579. Totnes Guildhall, built in 1553 and rebuilt in 1624, is another preserved by the town council with an early courtroom.

Rye and Stamford are two examples of Georgian town halls where the court shared a room also designed for council meetings and entertainments. Since the departure of the court (only recently in the case of Stamford) these have continued to function as they were designed, as flexible spaces for all kinds of community uses. The biggest change is the loss of the LCD's rent. Court fittings in such buildings are rarely extensive and sometimes movable (eg South Molton, Devon), because otherwise they get in the way of the room's other functions, and so the tensions between the desire to preserve court furniture and the need to adapt the space for its new life are less significant than in single-purpose courtrooms. The dais can stay against the wall, and benches, dock and witness and jury boxes possibly moved to the sides, if they are not already there.

During the nineteenth century town halls grew increasingly complex and splendid, especially in the new industrial towns and cities. Most, if not all, were built for use by Petty and Quarter Sessions and sometimes Assize Courts and were provided with separate courtrooms and support accommodation, and often separate entrances. Since these facilities have been mostly superseded by more spacious replacement buildings, many of the courtrooms have simply been left empty (for example at Leicester and Lancaster), tolerated where the building is not bursting at the seams because they make few additional maintenance demands. Some earn a useful income in the process as film and TV locations (such as St Albans), a lucrative sideline for a number of redundant courts, including the Assize Courts in St George's Hall, Liverpool and Kingston County Hall. Others, with the minimum of

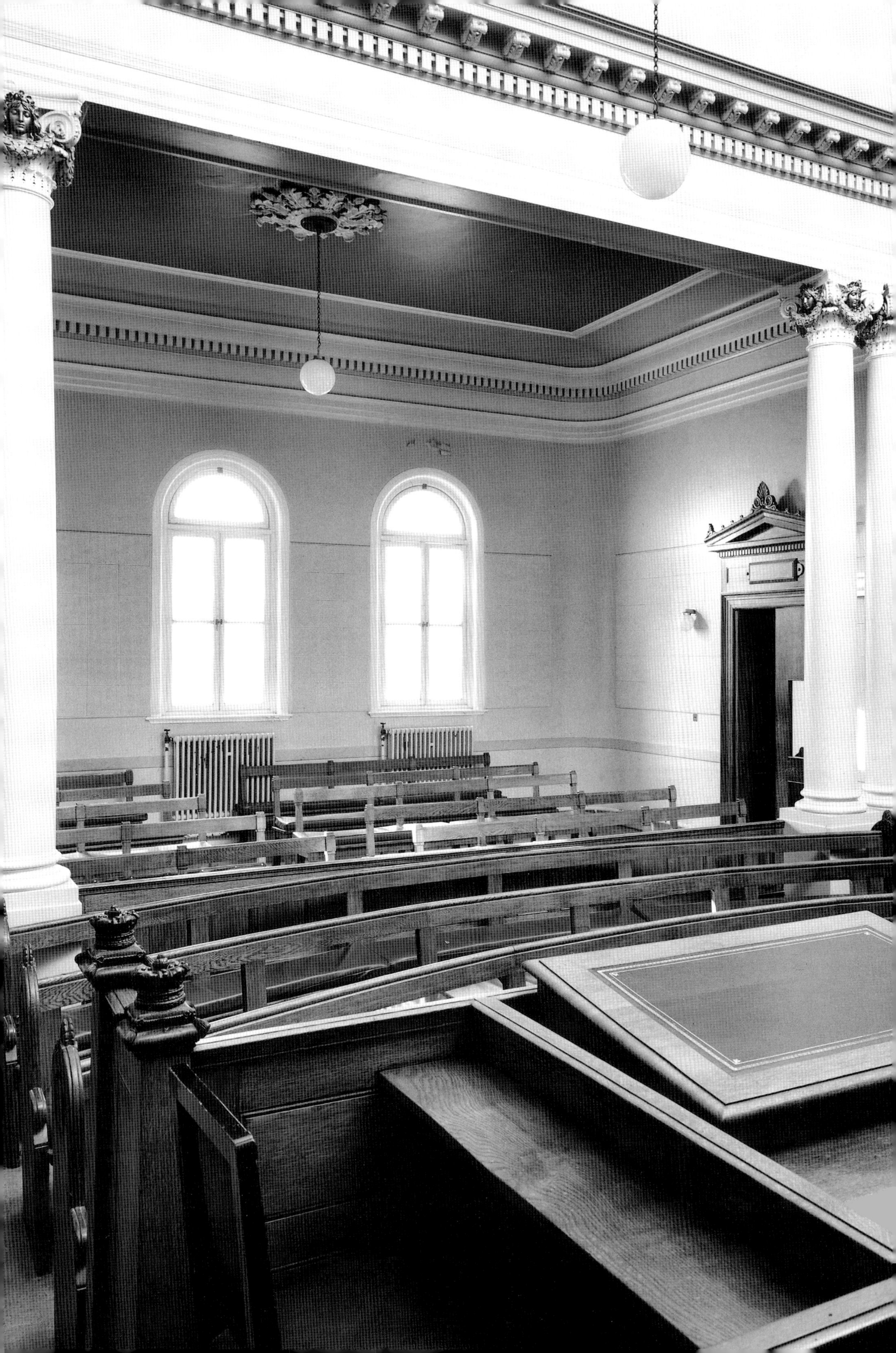

alteration, now serve as staff canteens (Middlesbrough) or smoking rooms (the Borough Court at Leeds Town Hall, where the redundant Assize Courts have been stripped). In other cases there has been pressure to make more intrusive use of these large vacant spaces. Kendal's Petty Sessions Court, for example, became the council chamber, losing the fittings between the bench and the public gallery to new seating and tables.

Victorian County Courts have proved particularly adaptable. County Courts were the first series of courthouses to be built by central government and under successive Surveyors to County Courts, Charles Reeves and Thomas Sorby, from 1846 to 1870, a definite house style emerged – dozens of buildings were erected to similar one or two storey Italianate designs. Well built and spacious with reasonably straightforward plans (designed without the need for complex segregated circulation and custody facilities) and very rarely with significant courtroom interiors (many have lost most of their original fittings), dozens have found new lives. The impressively wide variety of new uses – churches, temples, houses, flats, pubs, restaurants, offices, an art centre – is a testament to the flexibility of the building type. In larger town and city centres the imposing two storey models, often on prime sites, have proved particularly attractive to pub and bar chains (as at Nottingham, Bolton and Stockton). In smaller towns a number have become offices and, in at least two cases, houses.

Partly because their interiors are normally considered by conservation officers of less importance than their external contribution to the townscape, few of these conversions have been carried out with a great deal of sympathy or flair. The most challenging aspect of adaptation is the treatment of the double-height courtroom. Better proposals have tended to exploit the potential of this wonderful space, but in many schemes it has been destroyed by subdivision, either vertically or horizontally.

The extent to which nineteenth century County Courts are being made redundant, and then adapted – often brutally – for new uses, makes the conservation of at least one example in its original condition all the more important. Fortunately one of the most lavish is also one of the best, if not the best, preserved: Cheltenham. Its closure is now inevitable and the case for its conservation as a historic monument, one representative for hundreds of others, is very strong. In fact here is a building that should be vested with the National Trust, in line with its welcome recent policy of acquiring examples of all kinds of disparate building types, such as workhouses, textile mills, middle class villas, suburban semis and back-to-back housing. At first glance a representative of the more glamourous and more architecturally splendid Shire Halls and Assize Courts would seem more fitting, but unlike County Courts, a number of these buildings, and their smaller brethren, the Sessions House, have been conserved and are already open to the public as museums. The need for Trust intervention is simply not there. By contrast, without action at Cheltenham by the National Trust, English Heritage and others there is a strong chance that not a single County Court will be preserved for the nation. County Courts may not be the most glamourous element of the court system, but they are an important part of our legal history and they deserve this recognition.

A significant factor is ownership. County Courts are owned by central government. The Treasury expects them to be disposed of for the best price possible. The State has no desire to act as the long-term custodian for any of these once they are surplus to requirements. By contrast Assize Courts and Sessions Houses are owned by county (and some borough) councils, which were historically responsible for providing facilities for the quarter sessions and assizes. When these were abolished and replaced by the Crown Court responsibility for providing accommodation was transferred to the Lord Chancellor's Department, which made a contribution of 90% towards the cost of running and maintaining the Georgian and Victorian Sessions Houses and Assize Courts in which the Crown Court sat, but which remained in local authority ownership. Those Sessions Houses that became Magistrates' Courts received similar support. Thus, when these courts are closed, or moved to new combined centres, the biggest and most immediate problem faced by their owners is that of having, suddenly, to find 100% of the funds needed to maintain and repair them. The financial burden of closure falls squarely on the councils, not with the authorities that have made the decision.

Naturally these local authorities would in most circumstances like to dispose of such unwanted, obsolete liabilities as soon as possible, but the best Assize Courts and Shire Halls are highly specialized and architecturally

Leicester Castle.
The empty Nisi Prius Court, under the magnificent twelfth century roof of Robert le Bossu's Great Hall. The future of this extraordinary building is still uncertain.

important, frequently listed Grade I. As the most significant buildings erected by county councils until the end of the nineteenth century, the home of county administration and symbol of crown authority, they are lavished with ambitious facades, impressive interiors and high quality finishes and furnishings; as multi-purpose buildings designed to accommodate trial by jury and hold defendants in custody they may have complex plans. In such circumstances finding a sympathetic commercial solution is virtually impossible, and where it has been attempted the results have often been extremely damaging to the building. Derby Shire Hall deteriorated disturbingly whilst standing empty after the Crown Court moved out. Because it had become clear that there was no possibility of finding a suitable private buyer, it has finally been restored and extended as part of the Derbyshire Magistrates' Courts PFI scheme. Nottingham Shire Hall was sold for conversion into a hotel, but the scheme collapsed and it was bought by the City Council for restoration as a museum. Devizes Assize Court has been empty for over a decade since it was sold. Numerous proposals have come and gone and, though the court fittings have been stripped out, it is seemingly no nearer to finding a viable new use.

Thus most of the great redundant law courts are kept in public ownership. Councils such as Liverpool (St George's Hall and the County Sessions House), Cumbria (Carlisle Assize Courts), and Staffordshire (Stafford Shire Hall) have found themselves lumbered with vacant Grade I listed buildings which have often been badly maintained for years and so face a large and multi-million pound backlog of repairs, and for which future uses are very limited. Some, such as such Leicester Castle and Northampton Shire Hall, have effectively been mothballed because their cash-strapped owners cannot afford to repair and fit them up for public display and other compatible uses, such as seminar and conference use. Even after discounting the huge restoration bill, which for the most part inevitably the Lottery would be expected to pay, the on-going maintenance and management costs would barely be covered by the returns. Where councils have decided to refurbish, the desire to make more productive use of the building has inevitably focused on identifying internal spaces for adaptation to new uses, which in turn threatens to damage or destroy important interiors. The problem is the highly specialised plan of the Assize Court, which is typically dominated by three large volumes, the two courtrooms and a central or entrance hall. Together, these

Spilsby Sessions House.
Now a local theatre. A drama of a different k

take up much of the floor area; there maybe a handsome Grand Jury Room too, but otherwise only small cellular rooms. Options for extensions are normally extremely limited, this being a common cause for redundancy in the first place. Therefore, alternative use is dependent on adapting at least one of the principal spaces, and that logically should be one of the courtrooms. This is made easier where original fittings have been lost. At Bodmin Shire Hall, the Nisi Prius court has been restored for display and the Crown Court, which had lost original furnishings, substantially cleared to create a flexible performance and exhibition space. At Stafford's 1790s Shire Hall, Court 1 (largely mid-nineteenth century, but with some eighteenth century fittings) has been preserved intact whilst Court 2 has kept its bench, dock, witness stand, jury box and public gallery, but lost later furniture (put into storage) to create an arts workshop. However, at Liverpool's St George's Hall proposals to create a bar and restaurant as part of a Lottery-funded restoration project would have required the removal from one of the outstanding courtrooms of most of the original fittings. Thankfully, criticism of this scheme has resulted in the catering facilities being moved to a less sensitive location in the building. Similar debates are likely if schemes for the restoration and reuse of Leicester Castle and Northampton Shire Hall come forward, but the pressure should be resisted at the best courthouses, such as St George's Hall and Northampton, which ought to be preserved with all their principal spaces, including courtrooms, intact. Removing high quality, original furnishings from a courtroom should be considered in the same light as removing pews from the best church interiors. They are an integral element of the architecture, the very heart of the building and its purpose. Likewise, subdividing the principal volumes of a fine law court should be as unacceptable as partitioning the nave and chancel of most churches.

The biggest dilemmas come with the buildings on the next rung down, particularly the grander Sessions Houses, like Beverley, Morpeth, Spilsby and Wakefield. Here the conflict between finding a viable new use and conserving courtroom interiors is at its most acute because they normally lack two of the Assize Courts three principal spaces – one of the courtrooms and the hall. Therefore about the only means of introducing a new use is by clearing most of the fittings from the single courtroom which occupies most of the volume of the structure. Of the examples listed above, this has already happened at Morpeth and Spilsby, and has been sanctioned at Beverley and Wakefield. Will any be conserved with a courtroom intact? Once again ownership is a key factor. Unlike the best churches, which would be unburdened from the Church of England and vested in the Historic Churches Trust to be cared for by the State, the burden of preserving Sessions Houses, like Assize Courts, remains with cash-strapped councils.

There is a limit to the number of courthouses with interiors that ought not to be substantially altered and reused that can be opened as museums by local government. Unless in the future councils are compensated for this duty of care by the State, or buildings are taken into the custodianship of the National Trust, both of which seem unlikely, then this is another important reason why most of the best courthouses still in use, such as Warwick, Lancaster and York Shire Halls and Boston Sessions House, should remain so. Once again, that old conservation adage that the best use for an historic building is the one for which it was designed, holds true.

32

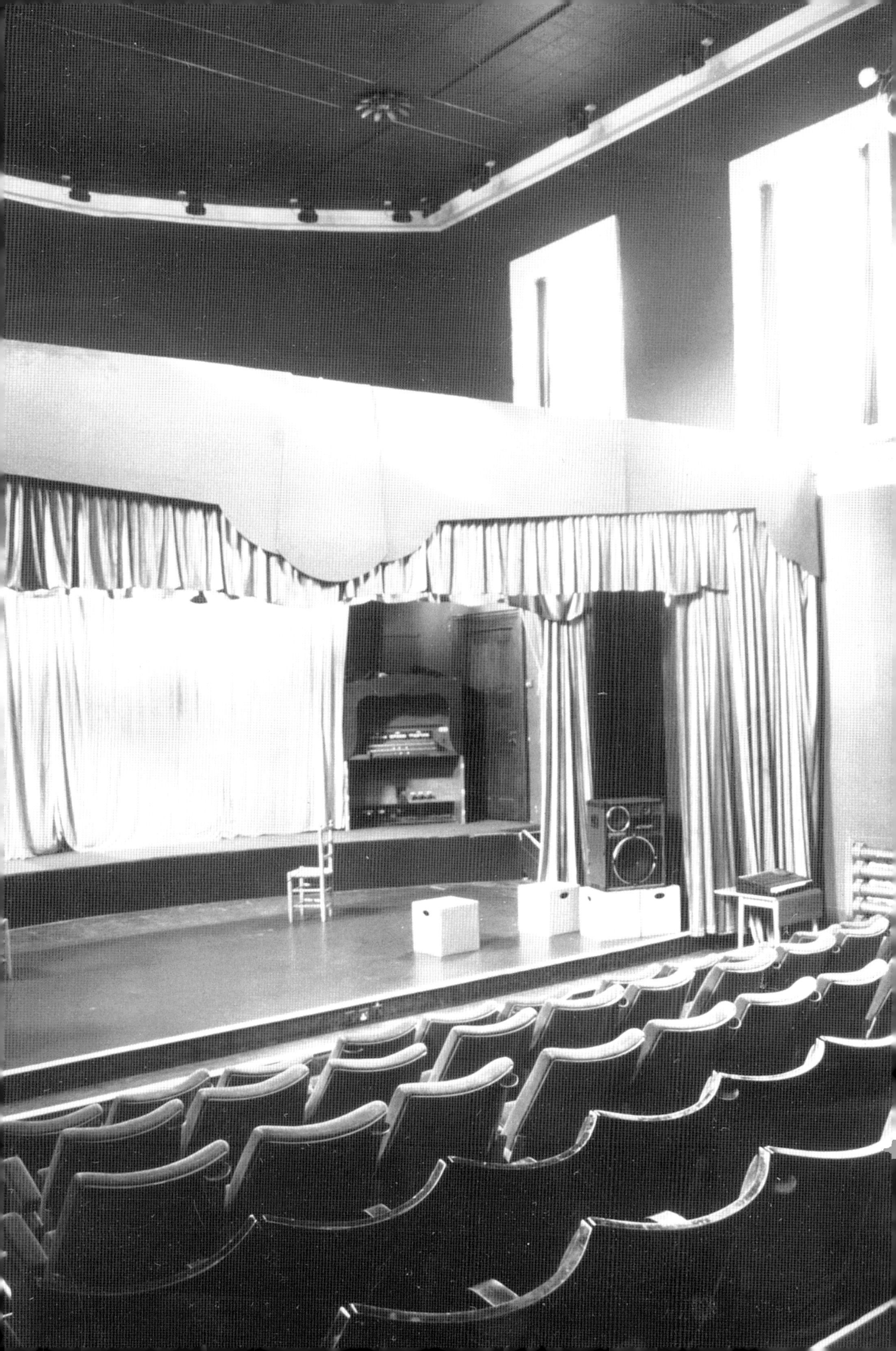

Conclusion

*The empty cells of John Dixon Butler's **Old Street Magistrates Court**. There is now an interesting scheme to reuse the building as a community law centre – a new public use for old public building.*

LAW COURTS are civic buildings, part of civic life and paid for by the public. There are numerous examples from all parts of the UK to demonstrate that where there is the will and the wit, they can continue to function efficiently and effectively as courts. They should only be discarded with the utmost caution and careful consideration because the best are incapable of adaptive reuse without damaging alteration to their interiors, and the alternative for them – preservation – is a burden that has already been borne by enough county councils and local authorities. No more should be made redundant. To that end the recent triumphs of local campaigners fighting to keep listed courts open in places like Knutsford and Kingston upon Thames are welcome reminders that closure plans can be overturned by determined resistance.

The climate has changed towards these buildings in the last decade. The LCD and Court Service have begun to grapple seriously with the responsibilities of their estate. Minshull Street, Chester and Worcester are all examples of the successful modernisations of listed law courts that have prevented the abandonment of these buildings. With English Heritage they are working to identify the most significant buildings in order to better inform policy toward them in the future. Together they have produced a set of guidelines for their conservation, alteration and management. Conservation Plans are being produced. At Derby the Court Service commendably forced PFI bidders to include the Shire Hall, and is doing the same with the wonderful Victoria Courts in Birmingham. For all this it deserves great credit. In addition, English Heritage is undertaking a detailed architectural study of the building type, which will considerably deepen our understanding of court architecture and undoubtedly result in beautiful and fascinating publications in due course.

Nevertheless, there is very little reason to believe that the age of closures is coming to an end, at least in England and Wales. PFI, IT and unceasing Treasury pressure will ensure that this is the case. The Government's response to the Auld Report on the reform of the criminal justice system, enshrined in the Courts Bill going through parliament at the moment, indicates a future that could see yet further rationalization of smaller courts. The Bill adopts Auld's recommendation that court administration should be unified by proposing a new executive agency to replace the Court Service and the forty two Magistrates' Courts Committees. Court Administration Councils will give local communities some say in the running of courts, with, it is claimed, 'real clout, in particular over decision… like the location of courthouses', though they will only be able to make recommendations: the final decisions will rest with the new agency that replaces the Lord Chancellor.

There is undoubted logic in unifying administration. The LCD press release accompanying the Bill says,

'Unification will allow many improvements to the way the courts work. For example, a unified courts estate will allow the heavy workload to be shared with another under-used courthouse, potentially saving one from closure while reducing delays at the other.'

Fine words. But one cannot help but imagine that the reality will be that rather than save courthouses from closure, the ability for example to transfer Magistrates' Court cases to Combined Court Centres will only hasten the demise of many existing law courts. The UK has a magnificent tradition of court architecture, but will its future be still more redundant than it is now?

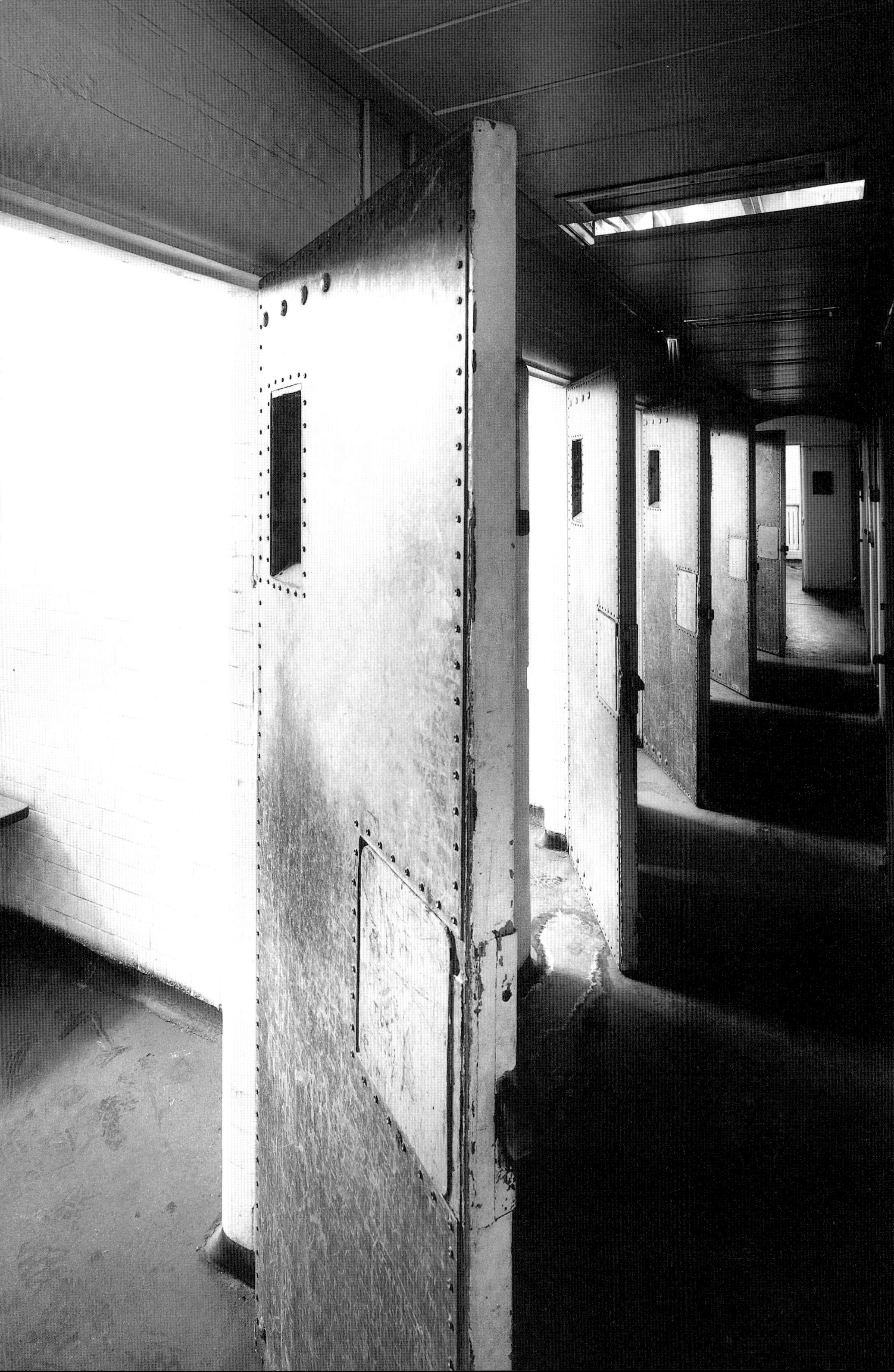

The history of law court architecture in England and Wales
The institutionalisation of the law
Clare Graham

THE idea of legal trial is older than the institutions which surround it, occupying a crucial position in almost every society. Social structures need to be supported by laws, but there will always be times when these are broken, or cannot be straightforwardly applied, and disputes arise. Resolving these grey areas into black or white must always be a fundamental priority for any government: if law and order are not maintained, any other kind of systematic activity becomes impossible.

This can be done by simply applying force, but a more peaceable alternative is to hold a trial, where the parties involved in the dispute submit their cases to the authorities for adjudication.

A trial is therefore by origin simply a problem-solving procedure, and in earlier times it had many more applications than is now the case. Bringing a problem to a judge's attention was the usual way of initiating many kinds of administrative activity in the days before extensive permanent bureaucracies existed. Our local government, in particular, remained entangled with our court system until well into the nineteenth century.

Trials began to need purpose-built accommodation when law became institutionalised: that is, when it acquired not only fixed methods of procedure but also written records and specialist staff. This process took place at a remarkably early date in England, being associated with the consolidation of royal government after the Norman Conquest. The solution of subjects' problems had always been both a religious duty and a matter of practical interest for the Christian monarch, but the king only truly became the fountain of justice when he acquired an efficient bureaucracy. By keeping written records he was, in the first place, safeguarding his own interests and administering his own possessions more effectively. But it also increasingly encouraged his subjects to consult him about their own problems, since his verdict could now be recorded in writing for posterity. He welcomed this development, not least because they were prepared to pay for the privilege.

This transformation of justice from a matter of duty into a source of profit provided the impetus for the institutionalisation of English law. The process really began during the reign of Henry II (1154-89), when the restoration of firm government after the anarchy of Stephen's reign encouraged a dramatic surge in property litigation. Compared to other parts of Europe, this was precocious, and it meant that England developed in isolation its own distinctive system of law, which became known as the common law. This began as the creation of practical civil servants, a spin-off from the royal administrative machine. If they wanted guidance in reaching a decision, they referred back to earlier precedents, usually in their own records. Common law interreacted very little with the Civil law (with a capital C), which emerged around the same time. This derived from Roman law and canon or ecclesiastical law, but was essentially the product of the new universities where it was studied and debated internationally. (It should not be confused with civil law with a small c: that is, the law used by individuals seeking to obtain private redress for their wrongs, as distinct from criminal law, which deals with offences against public order.) This academic product lies at the heart of most other European systems, including Scotland's. The Scottish legal system is and always has been quite distinct from the English one. On the other hand common law was introduced to Wales in 1536, though the Principality retained its own system of courts until 1830.

Many of the more confusing aspects of English legal history stem from these rather pragmatic origins. Common law was an inward-looking and self-perpetuating system, which rapidly became impossible to penetrate without professional assistance. As the specialised administrative staff attached to the courts proliferated, so did the lawyers who offered to steer a potential litigant through an ever more bewildering maze of paperwork and procedures. Professional training was entirely in-house until the nineteenth century, when common law finally became a university subject and both branches of the law brought themselves into line with other, newer professions by introducing qualifying examinations. Any legal system has a tendency to become culturally introverted and conservative, seeking to solve the cases presented to it by consulting existing laws and precedents, and appealing to the forces of tradition to uphold its authority. But England's isolation encouraged a legal culture which was to an even greater extent than usual both introspective and preoccupied by its own traditions, and many of the law's rituals provide visible evidence of this. Judges still conform to a dress code agreed in 1635; barristers still wear a black gown adopted as mourning for Charles II in 1685. Clearly the court setting also has a part to play in framing and ordering respect for justice and its traditions.

Historically, the common law represented a balance of centripetal and centrifugal forces. The former were focused initially on the person of the monarch as he travelled round the country, then on his high courts as these became institutionalised and settled down at Westminster. Again, this process began under Henry II (1154-89), and by the early fourteenth century Westminster Hall had become the usual home of the three high courts of the common law: Common Pleas, King's Bench and Exchequer. Each originally had its own area of interest, and its own officials, though distinctions blurred as the latter drummed up their fees by introducing legal fictions which increased their respective spheres of jurisdiction. The fee system also encouraged the Westminster courts to concentrate on potentially profitable business: in other words civil litigation, above all matters concerning landed property. The situation was complicated by the fact that until the mid-seventeenth century the monarch retained a prerogative to create other courts to supplement and sometimes circumvent the main system. Usually these were offshoots of the royal household: the Court of Chancery ultimately became a fourth high court, but other creations like the notorious Court of Star Chamber and the Court of Wards and Liveries [fig 1] proved shorter-lived.

Not everyone wanted to travel to Westminster to consult the high courts. Moreover, the maintenance of law and order also required the effective prosecution of criminal offences. Hence the system of assizes, which provided the centrifugal aspect of the common law. This emerged in the thirteenth century, and only finally disappeared in 1971: respect for tradition ensured a remarkable continuity both of basic organisation and of appearances over this long period, though other details naturally altered.

Fig. 1: The Court of Wards and Liveries at the Palace of Westminster.
Painting of circa 1585 at Goodwood House.

The counties of England were grouped into six circuits; twice a year, during the legal vacations, the twelve judges of the common law courts at Westminster divided into pairs and rode off around them. They held a variety of royal commissions, administrative as well as judicial: until the late seventeenth century assizes provided one of the main channels through which central government's policies flowed out into the country, and local information flowed back in return. On the judicial side, they could deal with both civil and criminal cases.

In each county the local sheriff was responsible for meeting and escorting the judges, finding them somewhere to stay and somewhere for the court to sit, and having the necessary people and documentation ready and waiting. Other officials responsible for maintaining law and order in the county would also be in attendance, most notably the justices of the peace (JPs, magistrates). These members of the county gentry gradually became the central government's chief agents of local administration. Also ready and waiting would be two sets of jurors: assize procedure required not only the trial or petty jury of twelve still familiar today, but also the larger grand jury of between twelve and twenty-three men, abolished in 1933. This presented any local problems or offences which had come to its knowledge to the judges; it also sifted through any criminal cases already before the court, dismissing without trial those where there was insufficient evidence to proceed.

As the main point of contact between central and local government, the assizes were accompanied by impressive ceremony. The sheriff had to find 'javelin-men' to keep order in the courtroom and escort the judges in procession, trumpeters to act as heralds, and a chaplain to preach at the assize service. Between assizes the JPs held four more meetings every year to settle outstanding judicial and administrative business. These quarter sessions were conducted in the same way as the assizes, with grand and petty juries, though with less ceremony. In addition, the monarch granted some cities and towns jurisdiction over their own affairs: they might have their own bench of JPs and sometimes their own quarter sessions and even a private assize. We should also note in passing that royal law was not the only law; other institutions, like the church, and individuals, such as lords of manors, held their own courts, sometimes in purpose-built accommodation [fig. 2].

Fig. 2: The Consistory Court in Chester Cathedral, 1636.

Fig. 3: The court of Kings Bench. Illuminated manuscript of the mid-fifteenth century.

The development of a specialised building type

Accommodation was initially makeshift, even for the high courts. Westminster Hall needed to be cleared out for other events like coronation feasts, and early inventories and a set of fifteenth-century illustrations show the courts' fittings originally amounted to no more than sets of tables, benches and bars [fig. 3]: even in the seventeenth century arrangements remained somewhat ramshackle [fig. 4]. In the counties and boroughs, responsibility for providing accommodation fell initially on the sheriff, then increasingly on JPs. Similar arrangements would be set up as required in the hall of the royal castle in the county town, where this was available, or in the town hall.

In the sixteenth and seventeenth centuries permanent court furnishings and specialised court buildings began to appear. To some extent this reflected increased business: from the 1560s, for instance, the two judges began to sit separately at each assize venue, dividing the civil and criminal business between them and forcing counties to provide a second court area. But new permanent settings also suggested an increased concern for the dignity of the court [figs. 1,2]. Their configuration followed earlier arrangements: judges and officials sat round a central table, the authority of the former usually being underlined by raised seating and/or a canopy, while suitors and their lawyers stood outside, at the bar of the court. Fittings could be simply placed within an existing space, so that while the inner area formed a carefully enclosed and hierarchically organised space for official deliberations, this remained an 'open court' with plenty of room for casual spectators [fig. 4]. But by the end of the sixteenth century special architectural features designed to help structure the trial process began to appear: separate routes up from the cells for prisoners, for instance, and private chambers for jury deliberations. Such arrangements seem to have been pioneered by corporations building new town halls, where spaces might double up as council chambers (see *Tittler*). But by the mid-seventeenth century county

Fig. 4: The south end of Westminster Hall with the Courts of the King's Bench and Chancery.
Anonymous drawing of about 1620.

ones were now consciously emphasised. The institutions of the common law, especially trial by jury, were presented as bulwarks of liberty and order, essential aspects of Englishness: historians may argue about whether official doctrines actually inspired general popular belief, but certainly their existence encouraged the dignification of architectural settings. Buildings used for trials were now treated as major architectural commissions, and for the first time their plans begin to be published, both individually and within volumes like Vitruvius Britannicus and its continuations. The county or shire hall features most frequently, as for example at Gandon's Nottingham Shire Hall, the seemly arrangement of its twin

importance of the common law. Though the administrative role of the assizes had diminished, its constitutional and also its social authorities were also beginning to build accommodation specifically designed to house legal ceremonies in some splendour. The new county hall at Northampton (1677-8) is the grandest example, built as an open L with separate court areas for the two assize judges at each end, and including a route up from the cells to the criminal court. Above, plaster ceilings added by Edward Goudge in 1684-8 celebrated the majesty of the law, with cherubs holding scales and swords of justice seated over the judges' heads [fig. 5].

Legal iconography grew further in popularity after the Glorious Revolution of 1688: motifs such as figures of Justice, swords, scales, fasces and staffs with caps of liberty suggest the growing ideological

Fig. 5: Plaster ceiling at Northampton Sessions House, (1677-8). Added by Edward Goudge in 1684-8.
Detail showing the area over the judge's bench in the criminal court.

assize courts and other accommodation offering a particularly exhilarating test of the architect's mastery of the symmetries of Palladian planning [figs. 6,7].

Court areas were now often distinguished architecturally by screens of columns, but effectively remained open, usually to a larger hall [fig.7]. By now the other basic requirements were recognised as another room where the grand jury could give criminal cases a preliminary hearing, and two more where the petty or trial juries could retire if necessary to consider their verdict. Some buildings were used for quarter sessions but not assizes, and here one court and one petty jury room sufficed. But since the court accommodation would still only be needed at intervals, it made sense to try to group other functions under the same roof. Together courts and hall made up a large space, suitable not only for county meetings and elections, but for new kinds of social events. Assizes and quarter sessions remained central to the calendar of local government but now they also became focal points within programmes of balls, assemblies and concerts. By putting up new

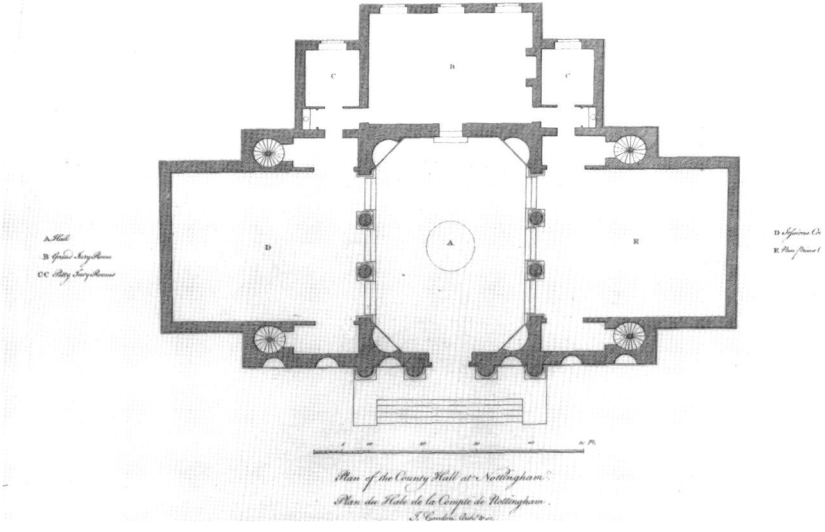

Fig. 6: James Gandon's proposed elevation for **Nottingham Shire Hall** (not as built).
John Woolfe and James Gandon, Vitruvius Britannicus, Volume V (1771).

Fig. 7. Plan of Nottingham Shire Hall as building (James Gandon, 1770-2).
As fig. 6

public buildings, provincial towns both advertised and fostered their intention to become centres for the diffusion of a new kind of polite culture. A distinct set of assembly rooms was the ideal, as at York, but others, like Salisbury, provided a county room within their shire or town hall.

By the end of the eighteenth century the traditional system of assizes and quarter sessions was coming under increasing strain. The growth in population, and the pressures of industrialisation and urbanisation meant that more court buildings were needed, some of them in new places. 1790 to 1840 was a period of intense building activity, during which virtually every county and borough had to rebuild or substantially alter its court accommodation to meet new and increasingly standardised criteria. For the first time it becomes possible to speak generally of the law court as a dedicated and specialised type: this development reflected not merely demographic and economic fluctuations, but wider social changes. New notions of 'propriety' expressed complex concerns: a new model of professionalism, an increased preoccupation with efficiency and security, half-conscious fears of disorder and pollution. Enclosure had always been the organising principle for the court furnishings themselves. But now an obsessive concern with categorising and separating users reached beyond the court area, establishing the most distinctive feature of the modern law court as a building type: the labyrinthine complexity of its circulation routes and service accommodation. The patterning of these private areas reinforced the distinctions suggested by the increasingly compartmentalised accommodation provided within the public space of the courtroom itself. While this desire to redraw lines between public and private activity was not altogether new, this period nevertheless saw dramatic changes, reflecting the 'lawyerization' of trial by jury. Representation by counsel, hitherto a rarity for criminal trials, was becoming much more usual. Procedures became more carefully controlled, more consciously rehearsed, and above all more longwinded. As the numbers of barristers travelling the circuits swelled, the etiquette governing relations with their clients, with the solicitors and even amongst themselves hardened. The judges (themselves recruited from the barristers' ranks) were also affected by these developments, and it is fair to say that the pressure to rebuild court accommodation within this period was usually generated by the legal professions.

Inside the courtroom itself, furnishings continued to grow in

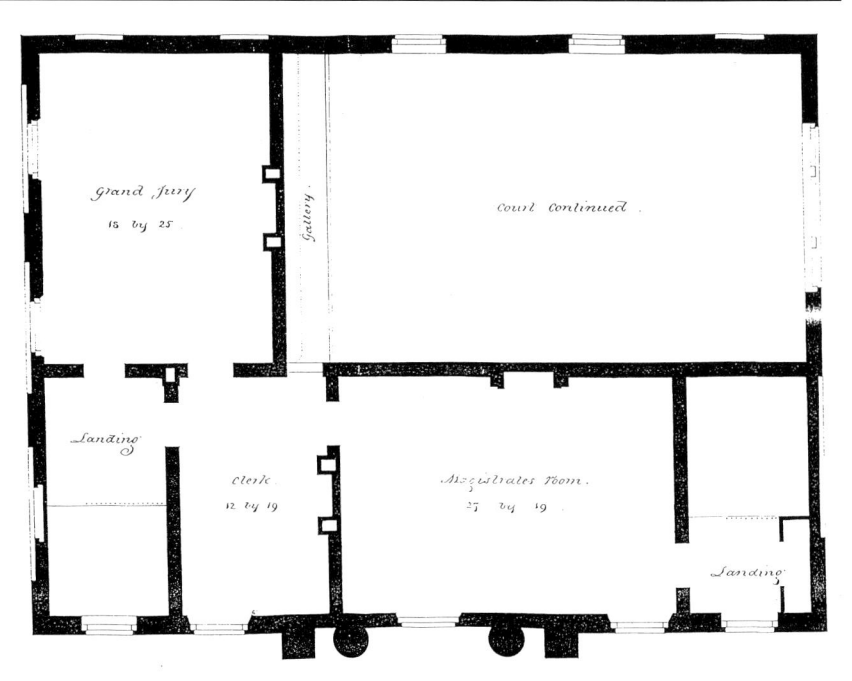

Fig. 8: Ground floor plan for Canterbury Sessions House. George Byfield, 1808

elaboration. One noticeable change reflected the lawyers' concern to maintain a professional distance from their new criminal clients. The 'bar' of the criminal court now became a box or 'dock' for the individual prisoner, again with a private subterranean route to the cells but now occupying an increasingly peripheral position within the court area [fig. 8]. Another important changeover, pioneered by Sir John Soane in some of the courtrooms in his new building for the high courts beside Westminster Hall (1821-6), was the provision of rows of seats and desks for the lawyers, which ultimately superseded the old idea of a large central table in the well of each court. But the really revolutionary changes took place outside the courtrooms themselves. Entrances and circulation routes multiplied, keeping different categories of participant separate outside the courtroom or at least controlling their encounters. Courts acquired an elaborate 'backstage': judges and barristers, in particular, as star performers, now demanded private spaces for robing and relaxation. The continuing system of commissioning buildings locally militated against the emergence of specialist court architects, but increasingly these projects demanded professionalism rather than inspiration: an individual who could not only grasp the complexities of legal etiquette, but build around them a court at once impressive and cheap enough to satisfy parsimonious committees of local JPs. Not surprisingly, the architect most successful in attracting these commissions was Sir Robert Smirke, with six English county halls and one Scottish to his credit – Maidstone, Carlisle, Gloucester, Hereford, Lincoln, Shrewsbury and Perth.

Lodgings

The new division between the formal exigencies of performance, and the collusive intimacies of backstage, stretched beyond the court building as increased attention was paid to assize judges' lodgings. Originally they had stayed in the royal castle where the courts were held, but by this date they had usually moved to an inn, or hired lodgings. However, during the first forty years of the nineteenth century at least a dozen local authorities decided to buy or build properties specifically for their accommodation. Judges' lodgings were usually built near to, or even as part of, the courts: at other times, and particularly during quarter sessions, JPs could use them to relax and transact non-judicial business. Those at Warwick, completed in 1816 were next door to the Shire Hall.

Legal reform and further architectural specialisation

By 1840 venues for assizes and quarter sessions as well as for the high courts had all largely been rebuilt, creating a new type of specialised building whose guiding principles still govern the courts of today. But the programme for rationalising and integrating the institutions of the law themselves, with all their idiosyncracies and inefficiencies, was only just beginning; it would not be complete until 1873-5, when the Judicature Acts reorganising the workings of the high courts themselves were passed. Soane's building had long proved inadequate and soon afterwards they were rehoused in G.E. Street's new Law Courts in the Strand (1874-82; see Brownlee). This was the flagship of the reformed system, an impressive and immensely complex building whose very existence helped attract an increasing percentage of legal business away from the provinces and into the centre of the system. But, for most people, the changes at a lower level were more noticeable. The reform programme created a two-way split in legal rituals, and this had the effect of generating further architectural subtypes. On the one hand, new courts were created at the bottom of the system, and used to process the bulk of legal work with little ceremony but as much despatch as possible. On the other hand, assizes and quarter sessions survived to grace the upper reaches of the legal system. Although their share of business dwindled, its symbolic importance meant these occasions were likely to be conducted with greater ceremony and housed in greater pomp than had ever been the case before.

Though the right to be tried by a jury of one's peers is still perceived as the tradition enshrined at the heart of the English legal system, the proportion of cases actually dealt with in this way has shrunk dramatically since the early nineteenth century. Today only those accused of a serious crime, or seeking substantial financial compensation in a civil court, are likely to have their case heard by a jury. This has been the inevitable outcome of the 'lawyerization' of the trial, a process which we saw had begun well before the 1840s. As the lawyers stretched out and complicated traditional procedures, trial by jury at assizes or quarter sessions became so slow and expensive that increasingly it had to be reserved for the most serious cases. Its survival and indeed its elaboration depended on the creation of an inferior network of local courts, which increasingly represented most people's first if not only point of contact with the legal system. These were the police courts (rechristened magistrates' courts in 1949) for criminal cases, and the County Courts for civil ones. Both systems were formally established in the 1840s, though they drew on earlier, more localised precedents. Procedures were intended to be relatively quick and cheap: these were courts of summary jurisdiction (i.e. without juries, though these were available for an extra charge at the County Courts), and at the outset at least the presence of counsel was not encouraged.

Police Courts

The police courts grew out of the additional petty sessions which JPs already held between quarter sessions, sitting in small groups to try minor cases, without a jury and not necessarily in public. In some urbanised areas, stipendiary magistrates were paid to provide this service more regularly at a public office. The creation of professional police forces (a development initiated in London with the foundation of the Metropolitan Police in 1829, but not finally imposed upon every county until 1856) provided an ever more constant stream of offenders for magistrates to deal with. Legislation passed in 1848-9 transformed these occasions into regular courts of law, with a formal and public procedure. Again, the burden of providing accommodation fell on local authorities: frequently, courts were built in association with a police station and/or town hall, to ensure that at least one of the three main classes of participant (police, prisoners, magistrates) would be conveniently to hand. In London, the courts were run by stipendiaries and attracted some central funding: Charles Reeves, who was the first Metropolitan Police Surveyor from 1843 until his death in 1866, built a group of new police stations, some with courts attached. In 1871 responsibility for the court buildings passed to the Office of Works, reverting to the Met in 1897. A distinctive group of nine new courts in Free Classical style was provided thereafter by the police surveyor John Dixon Butler: some of these, such as West London and Old Street, have been made redundant.

The County Courts were the civil equivalents of the police courts, again representing the legal reform programme's answer to the delay, inconvenience and expense increasingly associated with traditional trial procedure. The County Courts Act of 1846 established a national network of some 500 courts, divided into sixty circuits, each with a qualified barrister as judge. They were to hear pleas of personal actions where the debt or damage claimed was not more than £20. In many places he simply held courts in an existing building on his periodic visits, but elsewhere more permanent accommodation was needed. Central government broke with tradition by providing funding for this, and informally appointing Charles Reeves as County Court as well as Metropolitan Police Surveyor in 1847. Obituaries claim that he was responsible for sixty-four County Courts, a figure which seems to include alterations to existing buildings: following his death in 1866 his pupil and assistant Thomas Charles Sorby held the position until 1870, when responsibility passed to the Office of Works. Reeves and Sorby produced a coherent and interesting group of courts, mostly Italianate style; some drawings for them survive in the Public Record Office. Again, many have been made redundant.

Both police courts and County Courts had their own distinctive planning requirements, as did a third kind of new and relatively modest court building, the coroner's court. The coroner's office was an ancient one, but traditionally he had held inquests on suspicious deaths in inns or any other accommodation that was locally available. Only in the 1860s did changing ideas of propriety make this unacceptable. In some cases this led to the building of specialised courts, especially across the capital where the industry of the London County Council ensured that there was a network of no fewer than twenty-six venues in use in the 1920s. Until 1926 legal procedures required both coroner and jury to view the body at every inquest, and this meant that the courts were usually combined with mortuaries: many buildings followed the LCC's model plans, which were carefully drawn up to regulate the necessary contacts between the living and the dead [figs. 9, 10]. Elsewhere in the country inquests usually continued to be held in whatever accommodation was already available and today specialised coroners' courts remain relatively rare. Indeed even in London these buildings, with their bizarrely specific planning, largely became obsolete once inquest procedure changed and today only a handful remain in use.

Further up the legal system, in the more traditional setting preserved for a few cases, the second half of the nineteenth century saw formalities allowed to pile up, until trial by jury acquired an almost dream-like quality, suspended outside real time. This development was reflected during the later nineteenth century by the building of some very elaborate and indeed expensive accommodation, designed to enhance the ceremonial quality of these occasions. Such examples are not numerous, since after all in many places court accommodation had been rebuilt relatively recently, but both smaller buildings like the

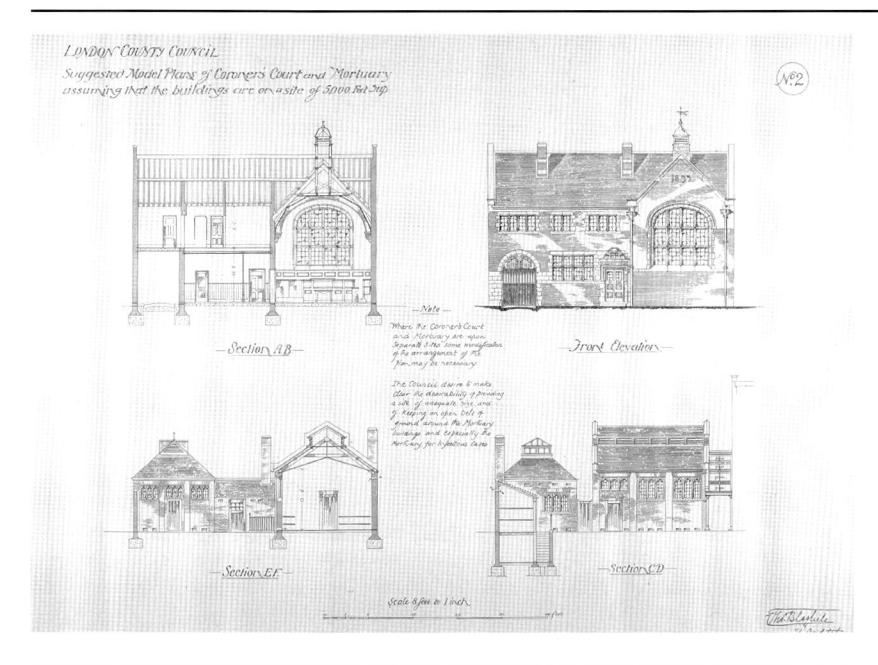

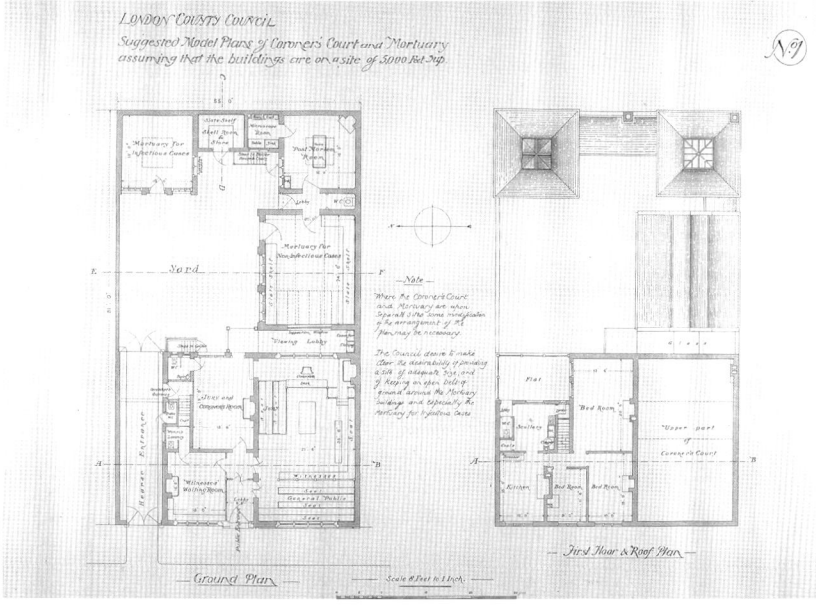

Figs. 9 & 10: Elevation, sections and plans for a model coroner's court and mortuary. Lithoprint, dated 1893 and signed by Thomas Blashill, LCC Architect.

Liverpool County Sessions House (Francis and George Holme, 1882-4) and larger ones like the Victoria Law Courts in Birmingham (Aston Webb and Ingress Bell, 1887-91) vividly demonstrate the symbolic importance which these events held for the local authorities which built them.

The decline of the trial?

In both larger and smaller court buildings, the period after 1840 saw demands for segregated circulation and service accommodation continue to grow; offices also expanded to accommodate a growing administrative staff. But while court buildings grew in overall size, courtrooms tended to shrink. The expansion of the popular press, and the subsequent growth in trial reporting, encouraged officialdom to assume that the professional journalist, rather than the general public, could uphold the principle of justice being seen to be done. This principle was formally recognised with the passing of the Children Act of 1908 which specifically excluded the public from the trials of those under sixteen, while permitting reporters to attend. Since then there have been further erosions of the principle of the open court – for instance restrictions on public access to domestic as well as juvenile courts, and on photography and indeed reporting for all trials. The effects of all this should not be exaggerated: the principle that trials should be open to the public remains generally applicable, and modern court buildings are still expected to provide some accommodation for the general spectator.

Nevertheless we would now rarely think of visiting a court building unless we have specific business there. The court system actually plays a much smaller role in our national consciousness than used to be the case, partly as a result of its own internal reforms. It was inevitable that many of the original applications of trial procedure would atrophy as the machinery of government grew; to survive in a more specialised age the courts had to provide a more specialised service. But it is also true that at certain points legal professionals deliberately restricted their role still further, in order to ensure their own reputation for specialist knowledge and impartiality. The period which followed the passing of the Judicature Acts in 1873-5,

45

in particular, was one of deliberate stagnation, during which judges shrank from involving the courts in new areas of legislation, leaving the field free for new administrative and commercial tribunals and arbitration procedures to develop. It is also true that, even within the legal system, the tendency to see the trial as the natural culmination of the legal process has diminished. In criminal procedure, the emphasis has effectively shifted to the period afterwards, when the sentence formally handed out by the judge will be monitored and modified by the authorities who carry it out. In civil procedure, the preliminaries have become more important: the elaboration and expense associated with 'lawyerized' procedure have increasingly made the formal trial a last resort, to be utilised only if participants cannot be persuaded to settle out of court.

All this is not to say that new court accommodation has not been, and is not still, needed. In this century the principles laid down by the nineteenth-century planners have actually been adhered to with remarkable tenacity. As Hanson's survey (see bibliography) notes, appearances have changed, especially in recent years: visible barriers and changes in level within the courtroom have been minimised, while decoration has become carefully neutral in an attempt to create a less authoritarian, more user-friendly, image for the law. So far as accommodation requirements are concerned there have been some new developments: for instance, the need to accommodate changing technologies, and to provide more rooms for consultations. But where segregation is concerned, requirements have if anything hardened still further. At a time when distinctions drawn on moral grounds in the early nineteenth century have been deliberately loosened up in many other building types, in this one they have been consciously preserved, usually in the name of security.

This is not the place to consider whether such a situation is necessary, or even desirable: what is clear is that it reflects the growth of the permanent bureaucracy attached to the court system, and the increasing role this plays in decisions about accommodation. This has created a greater insistence on standardisation and rationalisation, and ultimately a spate of redundancies for historic court buildings. In particular, patterns of provision were dramatically affected by the Courts Act of 1971. On the recommendation of a commission for legal reform presided over by Lord Beeching, assizes and quarter sessions was finally abolished, and replaced by a new system of Crown Courts, while responsibility for accommodation passed from the local authorities to the Lord Chancellor's Department, already responsible for the County and High Courts. Over the last twenty-five years its Courts Service has completed an extensive programme of rebuilding, often choosing to combine Crown and County Courts under one roof. Smaller local courts have closed as business has been concentrated in fewer and larger trial centres, organised so as to get the maximum number of 'court hours' out of accommodation. Usually it has opted for 'design and build' packages: while the results may work in practical terms, they have been criticised as visually bland and sometimes downright confused (see Pearman). On the other hand, there have been successful upgrades of older buildings (for instance at Chester and Manchester) which show that historic courts can be sensitively adapted to modern needs. It is hoped that lessons can be learned from this now that the focus has shifted to the magistrates' courts. Yet, as this report shows, outstanding historic court buildings are still being axed. Some redundancies maybe inevitable, but others can be improved and retained. Otherwise finding satisfactory alternative uses could prove problematic, especially for larger examples, despite the success stories outlined here. The elaborate court fittings and circulation routes which lend these buildings such distinction can also make them difficult to adapt to other purposes.

Bibliography

J.H. Baker
An Introduction to English legal history,
London: Butterworths, 1990 (3rd ed.)

David B. Brownlee
The Law Courts: the architecture of George Edmund Street, New York:
Architectural History Foundation, 1984

J.S. Cockburn
A History of English Assizes, 1558-1714,
Cambridge UP, 1972

Mark Girouard
The English Town,
New Haven and London:
Yale UP, 1990

Clare Graham
'Sudden death and the LCC: accommodation for inquests in London before the First World War',
arq, vol.i no.2 (winter 1995), pp.60-9

Clare Graham
The development of the law court as a building type in England before 1914,
University of Sheffield PhD thesis, 1997

Julienne Hanson
'The Architecture of Justice: iconography and space configuration in the English law court building',
arq, vol.i no.4 (summer 1996), pp.50-9

History of the King's Works,
(ed. Howard Colvin, London: HMSO, 1963-82

Hugh Pearman
'Court napping', Perspectives,
no.18 (Oct 1995), pp.28-31

Robert Tittler
Architecture and power: the town hall and the English urban community c.1500-1640,
Oxford: Clarendon Press, 1991

The Victorian Royal Coat of Arms at Bedford Shire Hall

Clare Graham's book, ***Ordering Law: the Architecture and Social History of the English Law Court to 1914*** is published by Ashgate.

Henry Littler's 1902 Sess House at Preston

It can be done: modernising historic law courts – an architect's perspective

Jim Stevenson DipArch RIBA ARIAS

COURT buildings are the architectural manifestation of the relationship between society and the judiciary, a physical representation of law and order, where not only must justice be done but must be seen to be done.

Ayr Sheriff Court. Robert Wallace's 1818 building. A new extension at the rear provides three new court rooms.

They have always been a powerful civic statement, forming an integral part of the historical infrastructure of our towns and cities; they are often landmark buildings in important civic spaces. It is therefore no surprise that a high proportion are now listed. However, in the last thirty years the need to accommodate the changing requirements of the modern court system have placed tremendous demands on these venerable structures. These requirements fall into three strands:

Physical

- The insistence on far greater segregation of parties. Up to five entirely separate circulation routes are now required to keep the judiciary, jury, defendants, public / counsel / court staff and special witnesses apart outside the courtroom. The legal process itself can be damaged if the wrong parties meet. Preventing this, whilst still creating simple logical plans, is a vital but complex task.
- Greatly enhanced security measures: society has become much more violent.
- New types of courts and trials – large fraud trials, terrorist trials, family courts and youth courts and the new Civil Trial Centres – each bringing new and different planning requirements and demands.
- The requirement for far more extensive ancillary facilities then ever before – crèches, witness support accommodation, interview rooms, better catering facilities: room for all these has to be found.
- The need to embrace the Disability Discrimination Act in buildings which by their very nature have many different levels, no more so than within the 'theatrically' laid out courtrooms themselves. Many law courts, civic buildings built as grand public gestures, have grand public entrances, which often means many steps. Nobody should be forced to use a dingy side entrance instead.

Technical

The impact of information technology. IT infrastructures in courtrooms, video links for remote witnesses and evidence, the digitization of court papers, the storage of central records and the computerisation of management and information in a court building, all place unforeseen demands on buildings designed one or two hundred years ago.

Environmental

The desire to de-institutionalise court buildings, to make them more user-friendly. Members of the public rarely use courts; when they do it is normally under conditions of some stress. The designer needs to make their visit as stress free as possible by

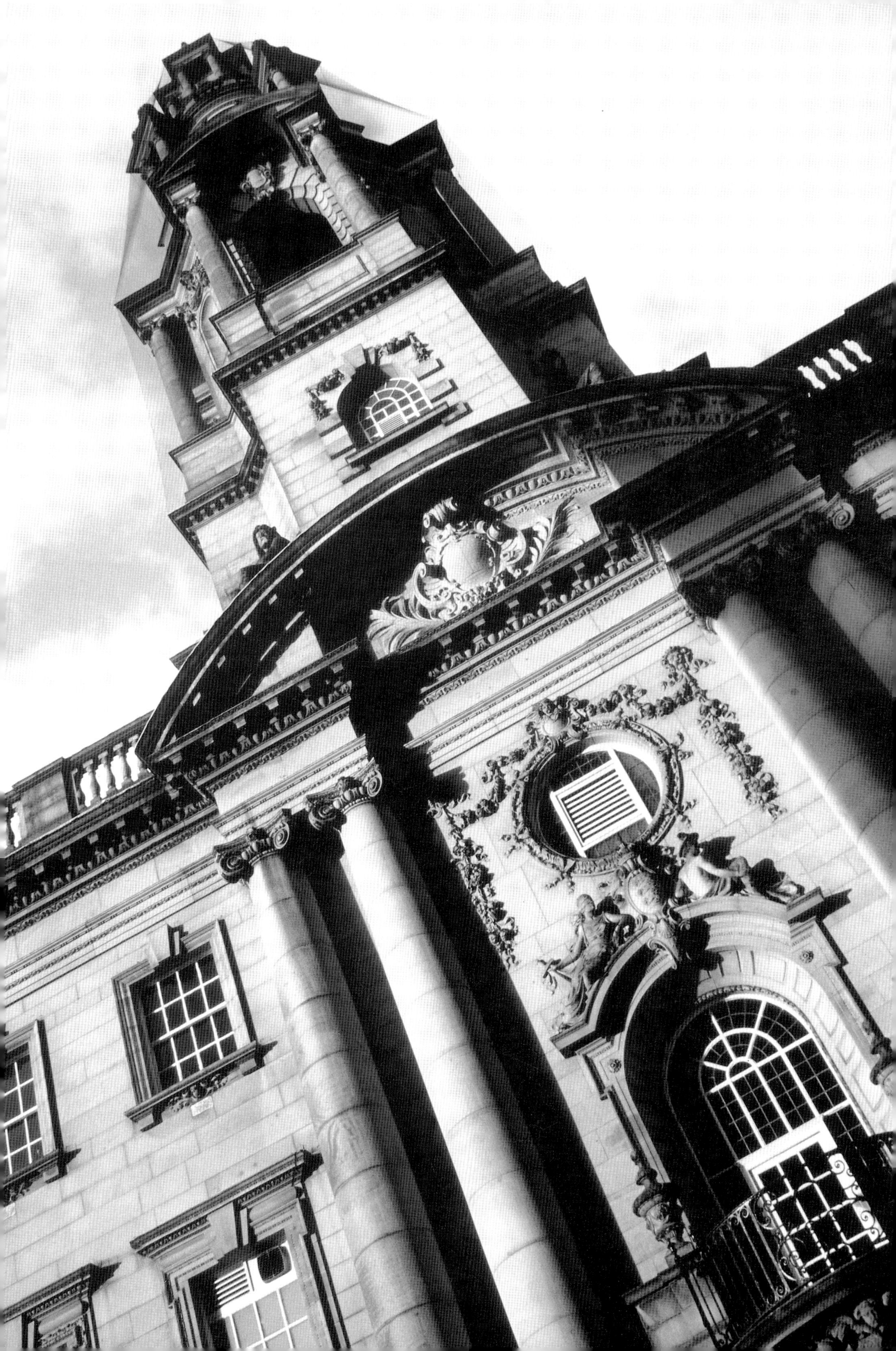

Minsull Street.
The central atrium, circulation segregated horizontally around it.

making a building as unfamiliar to them as a law court as logical – as 'readable' – as possible. Creating attractive, generous spaces, with the use of natural lighting and high ceilings for example, is an important means of reducing anxiety and conflict, which not only makes a building a more pleasant place for users and staff, but also contributes directly to the quality and efficiency of justice dispensed within it.

Beyond these pressures is the impact of the method by which new and upgraded court facilities are financed and managed. Virtually all court building programmes are now procured by the Private Finance Initiative, the partnership between the public and private sectors whereby the private sector provides the accommodation requirements, including its management and maintenance for twenty five years, for an annual payment. The facility management cost of a court building over this time scale far exceeds the capital cost of its construction. Thus, new buildings, designed in terms of 'life cycle costing' to be flexible, easy to run and maintenance free are what the private sector wants, and the efficiency, management and maintenance costs of existing buildings are put under intense scrutiny.

This all represents a considerable design challenge to architects attempting to secure the re-use of listed court buildings as part of PFI courts projects. The Lord Chancellor's Department, conscious of its obligations, has prepared Conservation Briefs for its significant listed buildings and has jointly with English Heritage published guidelines on historic court buildings which elaborate on national planning guidance (contained in PPG15). The presumption is that listed buildings should be re-used unless it can be proven that the retention of that building would detrimentally affect the operation of the court business. This creates challenges that require architects skilled in the adaptation of old buildings and with considerable experience of law court design.

But what exactly can historic law courts offer the modern judicial system? Well, for a start, spaces of a generosity and quality that could never be repeated today, and courtrooms that, though possibly in need of modification, create an environment and 'theatre' that is greatly valued by the Judiciary – they, certainly in the main, intone the idea of the Majesty of the Law. They are generally robustly built and planned and capable of absorbing a high degree of alteration. Despite the widespread belief that they are expensive to run, solid Victorian buildings, if maintained properly, are certainly less problematic than their 1960s and '70s cousins, constructed of immature and little-understood structural systems and materials, with building services of dubious efficiency.

The problem is that meeting the needs of modern law courts in terms of building services, circulation and accommodation frequently threatens to damage, or even destroy, the historic integrity of old courthouses. However, I believe that armed with a detailed understanding of the process and the client's needs, and with some 'interpretation' of the design requirements, a large number of these buildings can be retained and refurbished fit for efficient and effective use. A number of projects we have been involved with illustrate how this can be done.

One was the Crown Courts at Minshull Street, Manchester. The building was a Grade II Florentine Gothic police court, resplendent with towers, tourelles and gables, designed by Thomas Worthington and completed in 1885. It had been subject to fairly brutal modifications in the 1960s, and by the 1990s was 'closed for business'. The four original courtrooms were magnificent Victorian interiors reflecting a quite different process and indeed system, but the Lord Chancellor's Department's need now was for ten Crown Courts and the significant increase in ancillary accommodation that went with them. The key to the transformation was a relatively small but adjacent empty lot on which an extension could be built, and which, more importantly, allowed us to open up and reuse the original courtyard as a new entrance hall and central atrium, accommodating the different segregated users at different levels. Public facilities were provided on two levels, with old leaded-glass panels used to screen the jury circulation and opaque glass screen to veil the judges accommodation above. Four new courtrooms were formed out of redundant Victorian spaces and two new courtrooms, the main plant-rooms and open-plan office accommodation were located in the new building. Thus by turning the circulation 'inside out' the old building could be retained and upgraded.

A great deal of reorganisation of the historic interiors was required, but architectural elements were retained wherever possible and replicated where necessary in

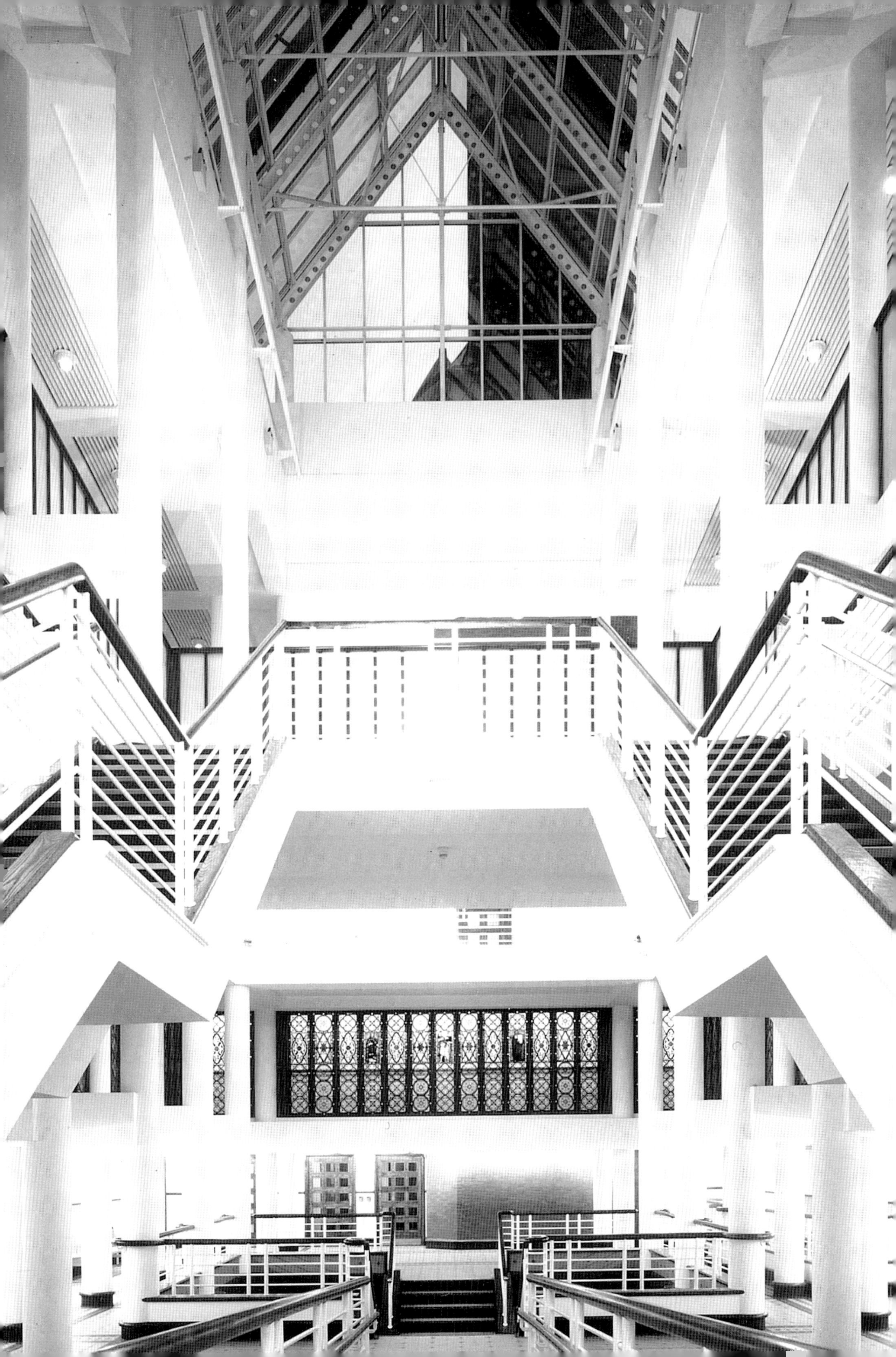

Preston Sessions House.
A glorious Edwardian interior

order to maintain the integrity of the old fabric, which is much loved by the Judiciary and by the staff who work there. The Victorian courtrooms were subject to subtle acoustic treatment and lighting, with extensive use made of the voids beneath the staged floors to introduce the new building services. Changes made to the docks to adapt them for multiple occupancy, and to the jury benches, were designed in keeping with the original.

The atrium is the heart of the building – a lively internal street environment where the day-to-day life of the court, the business conducted between the public, advocates, and witnesses, is played out. Without the available adjacent lot and the existing courtyard which enabled us to create the atrium, the future of this building as a law court would have been bleak.

At Preston the completion of a new Combined Court Centre enabled the Court Service to look afresh at the refurbishment of the 1902 Sessions House, a grade II listed Edwardian Baroque building in the town centre designed by the County Architect, Henry Littler, which housed the Crown Court. Typically, the original circulation was by now utterly inadequate, with insecure segregation of jurors and the public, support accommodation was lacking and building services were well below required standards. Later additions and alterations had been imposed unsympathetically on the building – and this time there was no adjacent lot on which to expand.

The solution was to concentrate trials requiring a higher level of security at the Sessions House, a smaller and self-contained facility. By removing a later courtroom and jury accommodation the building was made to function much more efficiently, and the transfer of some functions to the new court building created space for new support accommodation. Many of the interiors were restored to their original condition: Edwardian wall tiling was revealed; sound lobbies were added to the courtrooms allowing the tiled floors to be exposed where these had been carpeted to reduce noise; and the carpets and furniture in the original judge's library and dining room were renovated for their new use as committee rooms. The courtrooms were remodelled to rectify the circulation problems, with new doorways formed in the original panelling and stonework, and comfort cooling was added to overcome the serious overheating, using the voids under the staging. Appropriate signage was also installed.

At the Shire Hall potential redundancy and vacancy was avoided by working alongside a new facility to identify a niche use that reduced the demands on the building. This enabled it to be effectively and efficiently restored and modernised by a process of 'de-tuning', that is removing accumulated alterations and accretions.

As at Minshull Street, an extension forms part of our scheme for Ayr Sheriff Court, where we increased the number of courtrooms from two to five for the Scottish Courts Service. A Category A listed building by Robert Wallace of 1818, the courthouse has a grand portico and a domed hall (inspired by the Pantheon) containing a glorious sweeping staircase, and originally contained one sheriff's courtroom and a council chamber.

The solution here was to retain the magnificent 1818 building and its two courtrooms (one being the original council chamber) and replace the indifferent Victorian additions tacked on to the back with a new extension housing three additional courtrooms and support accommodation, thus providing the five courtrooms and all the ancillary facilities required, whilst at the same time undoing unsympathetic past alterations to the original building.

However, despite the success of the projects I have described, there are occasions, when for an accumulation of reasons it may

Minshull Street, showing the older part of the building to the left, built in 1871, and the 1996 extension to the right and the roof of the atrium.

Minshull Street, Manchester: Worthington's flamboyance

not be possible to retain and reuse an old court building. In Northern Ireland we have designed Belfast's new city centre court complex at Laganside for the Northern Ireland Court Service under the first court building PFI to reach financial closure in the UK. The new building, containing sixteen courtrooms, replaces a number of indifferent magistrates', county and crown courts buildings in and around the city. Sadly, one of these is the Category B-plus Crumlin Road courthouse of 1850, by Charles Lanyon. Though the cost and practicability of bringing together magistrates, county and crown courts at Crumlin Road did on their own raise significant problems, there were also unique difficulties associated with the location and the building's role in the history and politics of the Troubles; away from the unique circumstances of Northern Ireland it may have been possible to refurbish it to meet the design brief. For now the private service provider of the new combined centre is tied to maintaining the old building in a stable condition until, hopefully, with continued improvement to the political and economic climate in the Province, a viable long-term use can be found.

Another of our recent projects was Derby's Grade I listed Shire Hall, built in 1659 to the designs of George Eaton and extended in the 19th century, where we were engaged as historic buildings adviser to one of the ultimately unsuccessful consortia. The building forms part of a PFI project for the Derbyshire

Magistrates' Court Committee. A conservation brief was prepared by the Lord Chancellor's Department which required that the Shire Hall, empty since the departure of the Crown Court, to be reused as Derby's new Magistrates' Court.

Clearly, it would not be appropriate to alter the historic fabric radically: the solution was to make the seventeenth century building an entrance and circulation space for a large complex of new courts built on the rear of the site, an approach that works well in planning terms and is sympathetic to the fabric. In fact, it is a development of the historic role the Hall fulfilled after the courts themselves were moved out of it into new courtrooms in the 19th century. These two courtrooms could be used as hearing rooms, relieving the pressure for alterations to their well-preserved Victorian interiors. The real architectural challenge at Derby, therefore, was not one of planning, but one of the quality of the new architecture which will surround a Grade I listed building. There is one fundamental question I have not so far addressed: cost. But then it is very difficult to compare the full cost of adapting an old building to building anew. On the one hand at Preston's Sessions House a small amount of money was able to provide two fully-functioning courtrooms at a fraction of the cost of providing new ones. On the other hand, if an entire building has to be maintained operationally whilst alteration and additions are undertaken out of hours, the cost can be high. In addition, one cannot ignore the fact that refurbishing listed historic buildings may incur costs – such as the steeples to repair and the clocks and bells to return to working order at Minshull Street – that no new build project has to bear. But I would argue that these are relatively minor costs for the taxpayer to bear, and valid ones, because they are the means of ensuring that historic public buildings, important elements of our nation's heritage built at public expense, retain their central civic role within society for ours and future generations.

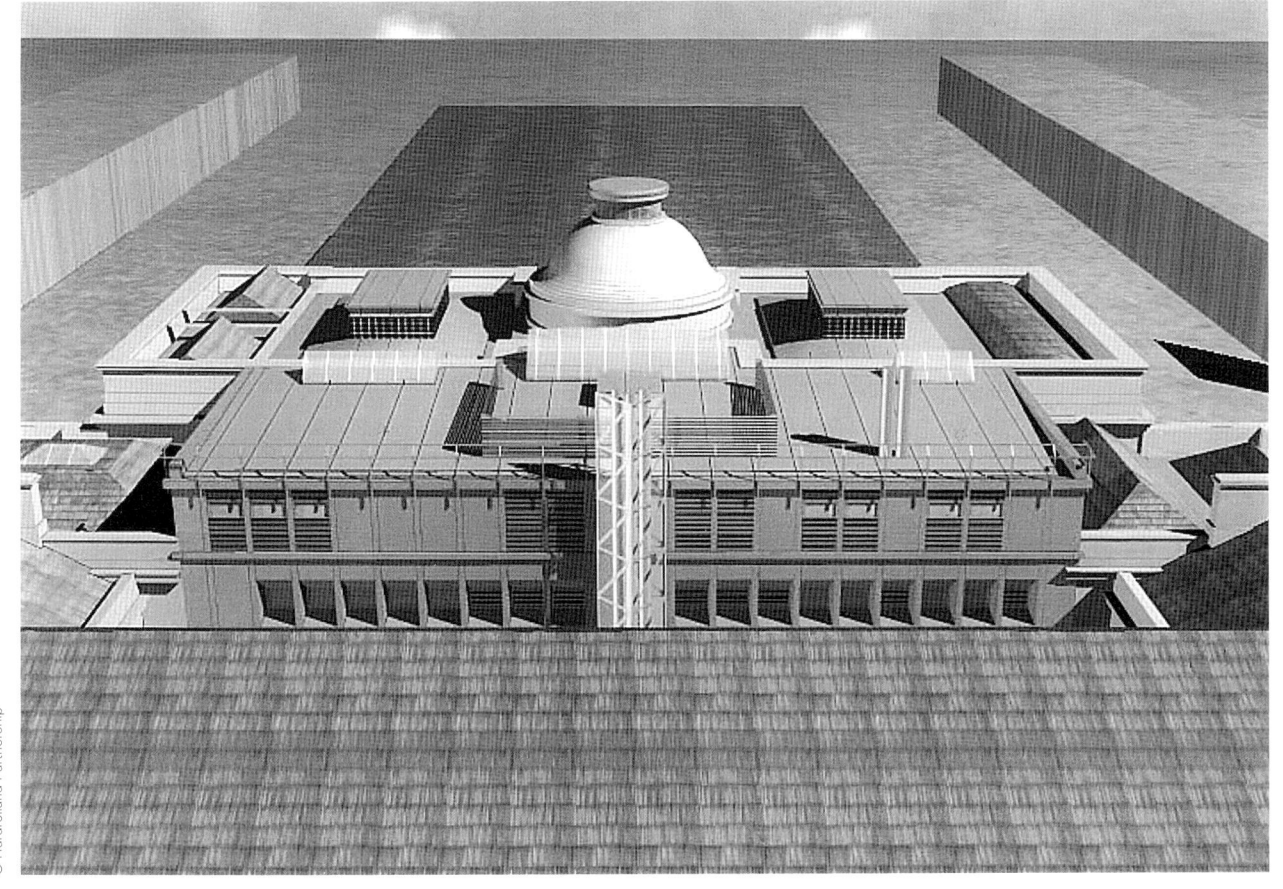

Ayr Sheriff Court.
The Hurdrolland Partnership scheme.

ENGLAND AND WALES

Appleby Shire Hall
The Sands   Grade II

As the Lord Chancellor's Department's massive Crown and County Courts rebuilding programme ran down, so the pressure to rationalize Magistrates' Courts was ramped up, and the gradual closure of courts and amalgamation of benches has been accelerated by the decision to reduce the number of Magistrates' Court Committees (the basic unit of administration) from 100 to around 40.

The results of this have been felt most sharply in rural districts where the closure of courts in small towns has deprived many rural communities of locally dispensed justice. In Cumbria, as discussed in the Introduction, four Magistrates' Courts closed in 2000 despite an appeal to the Lord Chancellor by the County Council. Appleby was one of these, and historically the most significant. The rather rundown building the magistrates sat in until the end was in fact the Shire Hall built in 1776-8 to house the Assizes in Westmoreland's county town. In this wild, poor and sparsely populated county far from fashionable society, financial means and architectural ambition were limited and compared with the splendour of Derby, York, Nottingham and many other Shire Halls you would be hard pressed to recognize this as a purpose-built law court and prominent civic building. Even Presteigne in remote Radnorshire had a courthouse of far greater architectural pretension (q.v.). It was designed by Daniel Bell of Whitehaven after plans supplied by Robert Adam were rejected as too costly. The principal façade, polite but simple vernacular of whitewashed rubble walls and unmoulded window surrounds, is little altered – except for the right-hand bay added in 1813 to provide a witness room and an office for the Clerk of the Peace. It lacks even the symmetry of the gaol built alongside six years earlier and still standing as the police station. The three tall ground floor windows light the County Hall (or Nisi Prius court) aligned parallel to the facade; the central door led directly into the public area at the back of that court (standing only - there is no evidence of seating); the three left-hand doors led directly into the back of the Crown Court, which was aligned at a right-angle to the frontage. Again there was no public seating, only a standing area behind the bar. This was justice stripped to its basics.

After the abolition of the Assizes the building housed the Magistrates' Court, which by the time it closed in 2000 was sitting once a week in only one of the courtrooms. Little remained of interest inside by then. The owner, the County Council, is now considering options for the reuse of the building which, fortunately, is not complicated by the physical relationship with the adjoining police station, from which it can easily be sealed. There is considerable local interest in reopening it as a museum or visitor attraction, and an action group has been formed with a view to becoming the new owner or tenant. A town museum has been suggested, and a feasibility study is currently underway to assess the case for a national Romany museum (Appleby has strong associations with the Romany community past and present through the famous Horse Fair). The idea has also been floated that the building could be adapted to house the town's new medical centre. At the time of writing the Shire Hall is still largely empty.

Barnard Castle Magistrates' Court
Queen Street, Barnard Castle, Co. Durham   Grade II

The Magistrates' Court in Barnard Castle closed in 1998, ending an 800 year tradition of court sittings in this unspoilt Pennine market town. Manorial courts had been held in Barnard Castle itself for centuries; Petty Sessions sat first in the Toll Booth, before the building was condemned in 1808 and the court was squeezed into the octagonal upper chamber of the pretty colonnaded Butter Market, with the jury box perched up in the rafters. In 1861 it moved again, to more suitable premises, when it joined the police station in two converted houses of 1827, now listed Grade II. An extension was built up against the rear and two courtrooms created.

The police moved out in 1977 and after the magistrates left the building it was bought by a local developer who has converted the building into six two bedroom apartments. The building stands in a quiet residential street, still very obviously a pair of terraced houses. The ashlar facade of local sandstone retains all its original sashes and the first floor is emphasized by subtly enriched window surrounds. The building is typical of a number of courts that have been converted from private dwellings.

Bedford Shire Hall

St Paul's Square

Grade II

Bedford is the current front line in the battle to save listed courts from the worst of the PPP/PFI process. The building is one of Alfred Waterhouse's rarer southern commissions, a severely red brick and terracotta composition, curiously Germanic in its Gothic. The interior is much more readily English – a hall with a simply detailed hammerbeam ceiling, lit by majestic perpendicular windows, which leads to two fine hammer-beamed court rooms retaining almost all their original fittings, and three other later courts of varyingly less interest.

The present Shire Hall replaced a building of the 1750s. This, typically, had two courts, one at either end of a large hall. In an evolutionary process similar to that at Derby, when this arrangement was found wanting in the nineteenth century (the Assize judges complained about conditions in 1877) two new courts were created behind the hall, designed by Waterhouse. However, no sooner were these completed than, in 1881, the town and county decided that facilities were still inadequate and Waterhouse was invited back to design a new front block to replace the eighteenth century Hall. This was completed two years later.

Waterhouse's building is already on its second life: after the Crown Court moved out it stood empty for a number of years before the Magistrates' Courts moved in. Despite the undoubted quality of the architecture and the completeness of the principal interiors, Bedford is marked by that unmistakably institutional feel of so many Victorian public buildings, built with such pride and care, but starved of investment for many years - quintessential 1960s light fittings, glossy cream and grey walls, dingily lit corridors, assorted furniture of many ages and origins and temporary partitions.

Undoubtedly the current facilities are inadequate. Upgrading is overdue and will be undertaken by PPP. The Bedfordshire Magistrates' Courts Committee are considering the bids of three consortia; it must decide between proposals to refurbish the Shire Hall or abandon it for another County Council owned site in the town.

The Shire Hall has a wonderful location on St Peter's Square in the centre of the town, bound to its rear by a riverside walk. The calming influence of such a setting on the highly charged experience of a court appearance should not be underestimated and it is no coincidence that a number of the new combined court centres, such as Nottingham and Newcastle, have been built alongside rivers. But PPP bidders are more concerned about the refurbishment and maintenance costs of a 120 year old listed building and the client has to be persuaded that the Shire Hall, so inadequate at the moment, can be transformed into an efficient and effective facility, and one that reflects their modernizing

© Martin Charles

aspirations. As ever, the challenge is as much about perception as anything else. There are now good examples of what the Shire Hall could become, prime amongst them being Minshull Street in Manchester. Rightly held up as one of the Lord Chancellor's Department's conservation successes, modernized and extended to the obvious satisfaction of its users, this piece of Gothic Victoriana should be an inspiration to the magistrates of Bedfordshire. To move elsewhere would not only be unnecessary, but would also be to condemn yet another large, specialized listed building to redundancy. There are already too many of these lying empty as evidence of the short-sightedness and narrow-mindedness of such a policy and it remains a truism that the best use for any historic building is the one it was designed for. After the debacle of Derby County Hall and the commendable determination of the Lord Chancellor's Department to return it to court use, the abandonment of Bedford would be a depressingly retrograde step.

Beverley Guildhall

Register Square   Grade I

Tucked behind the County Hall in Beverley is the austere Greek Doric façade of the Guildhall, put up in 1832-5 by Charles Mountain Junior. This screens a fascinating building which includes remnants of a fifteenth century house acquired in 1501 by the Keepers (the elected governors of the town) and a courtroom and council chamber of 1762-3 designed by William Middleton.

The magistrates last sat in the courtroom in 1985 and it has been used only for the odd civic occasion since then. Its principal attraction is Giuseppe Cortese's magnificent Rococo plaster ceiling. At the centre is a seated figure of Justice, possibly the finest allegorical device in any English court. Unusually she is not blindfolded, which does little for one's confidence in the standard of justice in Georgian Beverley. The use of decorative symbols of justice, such as figures of Justice or crossed fasces, was increasingly common in the eighteenth century, but still sparingly employed in England compared with continental courts such as The Tribunal in the Amsterdam Town Hall of 1640-62. Much more common was an expression of civic pride through county or city arms. The room also retains contemporary fittings, including Chippendale-style chairs on the dais. Since closure, only the tiered public benches inserted in 1827 have been removed, though the pillars that flanked these, originally from Hawksmoor's Beverley Minster galleries, remain.

Alarmingly, a large section of the cornice in the courtroom recently fell from the ceiling and the room was placed out of bounds. The cause was found to have been water penetration of the failed roof and lead valleys. Though the roof, with assistance from English Heritage, has been repaired, it is unforgivable that the council should have allowed such an important building to fall into such a state of disrepair in the first place. A programme of regular inspection and maintenance would have identified the problem at an early stage and reduced the cost – to the taxpayer – of repair. However, the opportunity is being taken to completely redecorate the courtroom after paint analysis revealed the original ceiling colours of off-whites and creams. The intention is to reopen the room to the public and explain through it how justice was administered in Beverley two hundred years ago.

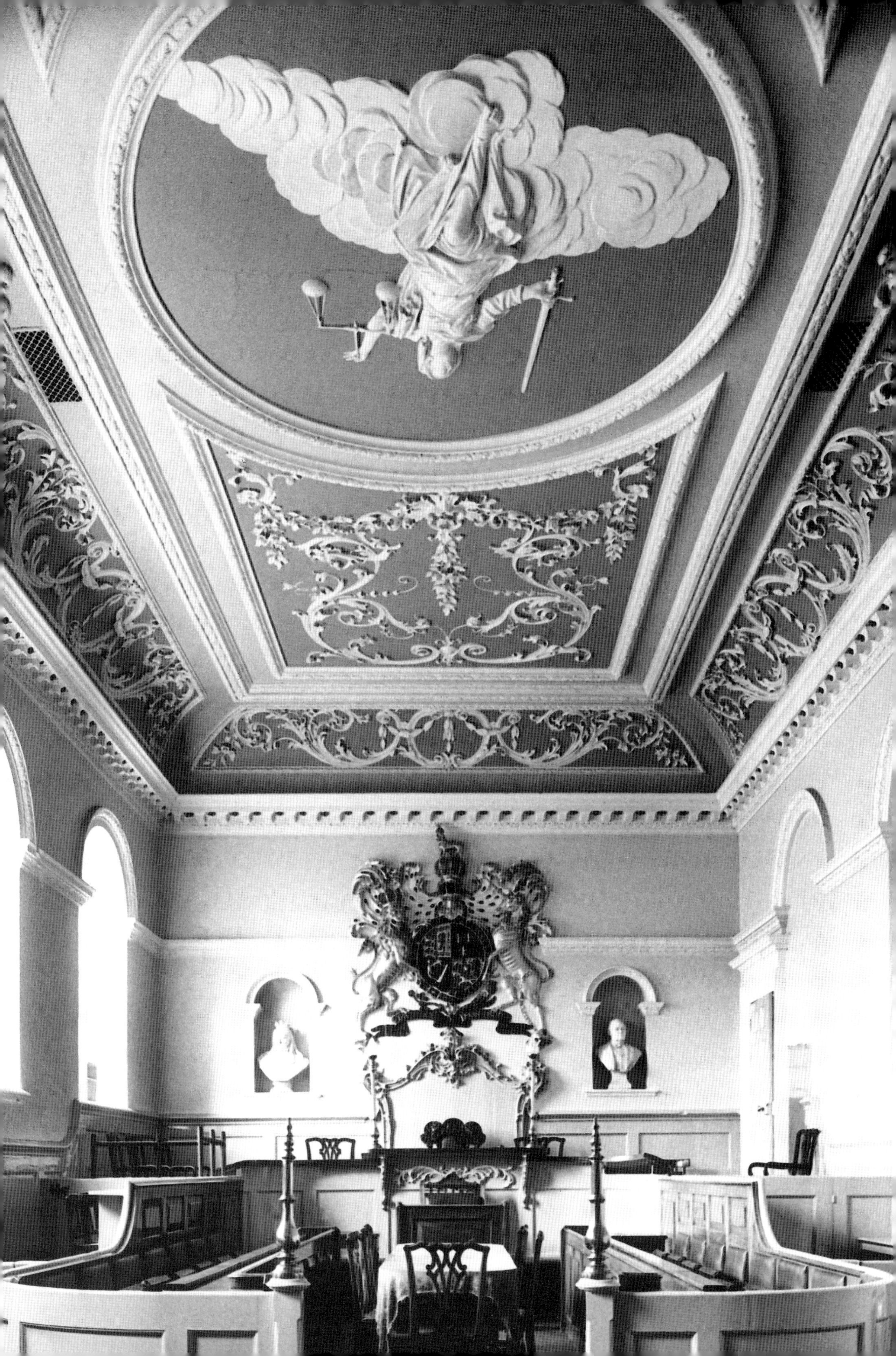

Beverley Sessions House

New Walk   Grade II*

Broad and elegant, New Walk is still the polite suburb no doubt envisaged by the city corporation when it laid out the street in the 1780s, with the financial support of the 'gentleman and ladies' of Beverley. Set back from it, and now partly screened by magnificent planting, is the handsome pale brick Sessions House of 1804-14, designed by Charles Watson of Wakefield, who went on to design another three Quarter Sessions courts in Yorkshire – at Wakefield, Sheffield and Pontefract. Only Sydney Smirke (with seven to his name) was a more prolific designer of courthouses at this time. Its planning is simple to comprehend. Behind the raised centre block and its Greek Ionic portico (crowned by a towering Justice) is the courtroom. The lower flanking wings contain all the support accommodation (with later extensions housing the police station). The courtroom, with a concave bench, retains its original fittings by Richard Jamieson of Beverley, including elegant Regency chairs (now in store). One room, almost certainly originally either the Grand Jury room or the magistrates, room was subsequently adapted as a second court.

The building was designed with an integrated house of correction, laid out along its central axis at the rear. In a remarkably early example of the kind of adaptive reuse SAVE has promoted for many years, this was converted by M L Whitton in 1880 into houses. Nearest the Sessions House lie the old men's cells, then the instantly recognizable central octagon that was the prison warden's house, and then the old treadmill block, now elegant housing in spacious lawns. The Sessions House housed the Crown Court until it was declared redundant and put on the market in 1999. It is still empty. It was bought by a developer who envisaged converting it into a restaurant, but the most recent application is for a hairdressing salon and spa. This would require the removal of all the seating and fittings from the courtroom (save the wainscot), and the insertion of a gallery.

Beverley is an excellent example of the particular problems faced when reusing a single courtroom courthouse. Much of the internal volume is taken up by the courtroom; the rest is small cellular accommodation. However, the courtroom, complete with its original furnishings, is also by far the most architecturally significant interior. Therefore, the commercial value of the building and the historic and architectural value of the building are in conflict in this focal space in a way often experienced with redundant churches. The latest application illustrates how difficult it can be to resolve the conflict. An intensive new use normally requires the removal of fittings and subdivision of space which is, from a conservation perspective, unacceptable. However, a more sympathetic low intensity use is frequently considered commercially unviable by developers or businesses. Thus single courtroom buildings are vulnerable in ways that multi-court room complexes, where an argument can often be made for the sacrificial conversion of one if another is conserved as found, are not.

A splendidly elaborate cast-iron Victorian urinal (listed Grade II), which once stood outside the courthouse, has been put into safe keeping by the council.

Birkenhead Sessions House
Chester Street — Grade II

Birkenhead is one of the most interesting but least known towns in the country. It was founded as an industrial speculation in the 1820s by the shipbuilder William Laird who hoped to share in the spectacular growth of Liverpool, on the opposite shore of the Mersey, and was laid out on a grid by James Gillespie Graham, a fellow Scot. At its heart is Graham's Hamilton Square, arguably the most spectacular in England, built up on a vast scale in 1825-44 with grand, four-storey sandstone terraces, book-ended with projecting pavilions and instantly reminiscent of the architect's work in Edinburgh. But the fledgling town suffered a severe depression in the middle of the century, linked to the failure of the first incompetent scheme to develop docks there and many of the projected streets were not built. When the economy recovered two ambitious civic buildings were erected. One was the imposing town hall built on the East side of Hamilton Square by C O Ellison & Son in 1883-7. The other was on the block behind, the 1885-7 Sessions House and Police Station, by T D Barry & Son. This, now the Magistrates' Court, has an equally sumptuous freely-classical exterior, easily a match for the Sessions House completed by the town's older rival across the Mersey in the year before construction was begun. The interior, sadly, is not up to the same standard, and has been much bashed about. The main courtrooms are on the first floor, now reached by a side door since the main entrance and stair on Chester Street was closed. The principal courtroom is bang in the centre, treated as a separate block rising above lower side wings, but it has been divided in two, with a suspended ceiling obscuring the top-lighting in the curtailed court. Oak fittings remain. What was the back of the court is now a waiting room. There are a further four courts, one with mid-twentieth century panelling. The public areas are a dismal shambles with vestigial remains of their original dignity.

The local Magistrates' Courts Committee is fully aware of these shortcomings and is planning some improvements in the short-term, but more fundamental, longer-term investment is dependent on the outcome of a Merseyside-wide PFI scheme. No doubt the array of exterior columns and statuary will give bidding consortia cold feet, but providing the number of courtrooms required is not demonstrably too great for the site, there should little excuse for abandoning the building. It occupies an excellent central location, between the town centre and the redeveloped river front, minutes from the train station. Solicitors, the police and other court users are clustered around. It is large and, as the police have left, the whole building is now available for expanded court facilities. The interior contains plenty of potentially attractive spaces and the general lack of surviving fittings affords designers greater flexibility when reordering it. At a time when so much effort is being made to revitalize the town centre and such large sums are being invested in restoring the nineteenth century terraces around it, it would be an act of civic perversity to vacate the Sessions House.

63

Victoria Law Courts, Birmingham
Corporation Street Grade I

With the exception of the GE Street's Royal Courts of Justice, this is arguably the finest court building in England, and certainly the most exuberant. This superb expression of Birmingham's unique reforming civic pride, known as 'the civic gospel', was completed in 1891 to the designs of Aston Webb and Ingress Bell, who beat 125 other architects in a competition. The city was advised by Alfred Waterhouse (architect of the widely praised Manchester Assize Courts, destroyed in World War Two), who provided entrants with a sketch plan for guidance. For the cost of £113,000 the city got a building more than worthy of its elevation to Assize town in 1884, and a fine Golden Jubilee Memorial to Victoria to boot. Terracotta is the key, red Ruabon on the exterior and buff Tamworth on the inside. This was moulded with almost absurd richness by William Aumonier – plain wall surfaces appear to have been outlawed. The style is eclectic – free Perpendicular cum French Flamboyant cum 16th century Flemish. Sculpture was by Aumonier, W S Frith and Harry Bates, who created the fine Victoria in the entrance gable. Not surprisingly, this terracotta fantasy was widely emulated in Birmingham; look no further than William Henman's General Hospital next door.

The principal interior space is the spectacular Great Hall, aligned across the street front, crowned by a magnificent hammerbeam roof and lit by five bronze electoliers and stained glass windows depicting Victoria's visits to Birmingham and the city's worthies. Originally this sumptuous entrance hall was reserved for the legal professionals, as the main element of a grand processional route that led down the Judges' Corridor to the exquisite domed central hall and the Assize courtrooms. The route terminates in another fine room, the Bar Library. The public entered the courtrooms from passages along the sides of the building. In our more egalitarian age all can enter as once only the lawyers did. The courtrooms are not as spectacular as the public spaces, but are nevertheless richly finished and furnished in generally early Tudor fashion. Court No.5 retains its plaster ceiling of pendant bosses.

As built, the Victoria Courts was a kind of proto-Combined Court, conceived to house not only the Assizes and Quarter Sessions, but the Petty Sessions and Coroners Court as well, in six courtrooms. The Crown Court moved out in 1988, since when the building has been exclusively a Magistrates' Court. There are now twenty six courtrooms in all, spread through buildings across Newton Street including the former County Court of 1882 by James Williams. For many years these facilities, which have accumulated haphazardly over decades, have been recognized as wholly inadequate and a PPP scheme for their complete overhaul is in the pipeline. The Lord Chancellor's Department, fully aware of the importance of the Victoria Courts, is committed to keeping these in the building, but with such extensive accommodation required this will represent a considerable challenge. The key in most instances to the successful modernization of historic court buildings is the availability of land for expansion. The Newton Street site can provide that space for the Victoria Courts, but the complexity of the split site, and the need to retain the Grade II County Court building, will require an approach of ingenuity and imagination. Nevertheless continued court use makes a great deal of sense, not only to secure the future of a Grade I listed building, but also because the site is in the heart of the city's legal district, with the new Crown and County Courts and many legal firms close by. The Courts Service's determination that Derby County Hall should house the town's new PPP-provided Magistrates' Court hopefully points the way towards its policy for the Victoria Courts, though the design quality of the adopted solution will need to be of a far superior kind.

COURT N°1

Bodmin Shire Hall

Mount Folly Square

Grade II*

Eventually, with the assistance of Lottery largesse, a success story. Bodmin is another of the breed of purpose-built courthouses that were put up at the end of the eighteenth century and beginning of the nineteenth in response to the pressure placed on the judicial system by the rapidly growing population.

The building is the work of Henry Burt of Launceston and was erected in 1837-38 to house the Assizes. It is a classic example of its kind. The handsome granite façade clearly articulates the uses within: to either side are the tall sashes lighting the two full height courts – the Crown Court to the left, the Nisi Prius to the right; between, under the pediment, are the windows lighting the upper hall from which the public galleries are reached; below are gated arches. Through these is the open entrance hall, and between a loggia at the rear rises a magnificent cantilevered imperial stair.

After the abolition of the Assizes the building housed the County Court until that retreated to Truro in 1987. Then it stood empty for four years until it was sold and converted into a restaurant. This quickly failed, and in 1994 it was bought by the Town Council. Initially the council planned to convert it to offices, retaining one of the courtrooms, but realizing that it could not create sufficient office space it drew up alternative plans – only made feasible by the creation of the Lottery – to restore the building as a living museum, a performance space, tourist information centre and local museum. With £800,000 from the Heritage Lottery Fund and a grant from the European Development Fund the project was given the go ahead and the completed centre was opened by the Queen in June 2000.

It has been an exemplary project. The external restoration has been matched by the repaving in granite of the town square above which the building stands (it was previously a tarmac car park). Internally, the alterations made to turn the Judge's Robing Rooms into kitchens were undone and the remains of the public seating (simple open-backed benches) in the courtrooms, which had been butchered to seat the restaurant diners, were assembled to fully equip the Nisi Prius Court. With all its other original court fittings still in place this has been restored to its 1850s appearance. Paint analysis revealed that the ceiling had been an off-white and the walls a warmer white distemper and that all the deal joinery, including the window frames, had been grained to look like oak. This decorative scheme has been beautifully restored. In the Crown Court the fittings in the well of the court have been retained, but the rest of the room has been cleared to create a display/performance area.

The project has revealed the ingeniously simple use of two consecutive iron screens beneath the grand stairs to segregate the public, the advocates, witness, jury and the defendants. The latter were brought up to the courtrooms from the 30 perfectly preserved holding cells in the basement. These are frighteningly small – little more than cupboards as wide as a man and two foot deep. They were of course only meant to hold a man for a day. Those awaiting trial were at least afforded ventilation; air was drawn up from the basement through cavity walls and into the courtrooms, to be extracted through pierced central roses of acanthus leaves.

The overriding impression is one of severity. Finishes are simple, and the austere entrance hall and stair – open to the elements – are bare ashlar. The character of the grey granite of course has much to do with this, but the sense of severity speaks also of the purpose of the building and the desire for those who commissioned it not to waste public funds on unnecessary embellishment – convenience and economy are the guiding principles. Nevertheless, the atmosphere of the place seems irresistibly appropriate for a building placed on the edge of the infamous Bodmin Moor.

Bolton County Court

Mawdsley Road

Grade II

Bolton County Court is a very handsome, but not uncommon building. It is a better than average example of the typical County Court built up and down the country in an Italianate style by Charles Reeves, Surveyor of County Courts, and his successor, Thomas Sorby. This particular example, of 1869, is by the latter. Its recent history is also typical: closure in 1992, vacancy and then conversion into a bar and restaurant. Middlesbrough (built 1904), Stockton-on-Tees (1860), Nottingham (1871), Westminster (1908), Dartford (1860), Huddersfield (1820) and Barnsley (1861) are all listed former County Courts that have undergone similar conversions.

In each case the attraction for the new owner was a substantial, impressive-looking, centrally located building – something with a bit of class; in each case (perhaps with the exception of the Westminster courthouse in St Martin's Lane, now a 'Browns' restaurant) little remains inside to identify the building as a court building; and in each case such alterations have been sanctioned because the interiors were considered of little significance, or substantially altered. Thus, at Bolton, the ground floor, which once was offices, has been opened up into a large uninterrupted bar, indistinguishable from hundreds of others up and down the county. The first floor courtroom is not even used, except for storage (the ground floor is big enough).

Many redundant County Courts are substantial, dignified buildings that form important parts of the streetscape in many towns. Fortunately, because they often occupy valuable town centre sites and can be converted without too much fuss, many have found viable new uses. But this has often been at the expense of the original internal plan and fittings, something that makes the case for the preservation of Sorby's Cheltenham County Court, with its complete and high quality interior all the more important.

Bolton Magistrates' Court
Le Mans Crescent   Grade II

The Magistrates' Court in Bolton is integrated with the police station in one quadrant of the Municipal Buildings, a crescent by local architects Bradshaw, Gass & Hope, completed in 1939: an ambitious piece of town planning that forms a grandiose backdrop to the town's magnificent Town Hall. The style is unreformed and confidently classical.

The main entrance is at the right-hand end, in a recess splendidly framed by two giant Ionic columns in antis. Inside a semicircular entrance hall leads to the left, to the magistrates' accommodation and the police station (and now to Courts 3-6, created from offices and other ancillary spaces) and ahead to the two original courts. Court 1 is grand enough to be an Assize court. The ceiling rises high into a pendentive dome and a large domed sky light. The public sit in a generous raised gallery at the back of the court. Court 2 is toplit as well, but smaller with a flat ceiling. Both courts retain their original, high quality, oak and bronze fittings and furnishings, as indeed do all the public spaces.

In 2001 the court was threatened with closure. The Greater Manchester Magistrates' Court Committee proposed to transfer Bolton business to Bury, claiming that the Le Mans Crescent building was 'inadequate' and required £700,000 spent on it over ten years to bring it up to an acceptable standard. The local reaction was furious, with widespread anger and disgust that a borough with over 250,000 residents could lose one of its core civic institutions, and the council and local law society pledged to do everything they could to keep the court open. The GMMCC defended its decision by saying that the move to Bury would – improbably – benefit court users in the future, reduce over-capacity and, more significantly, produce financial savings of around £350,000 over five years.

Over 20,000 people signed a petition organised by the local paper, the Evening News, which was delighted to announce in May 2002 that the GMMCC had backed down: Bolton would keep its Magistrates' Court, a victory it would seem for community action and common sense. However, this has not been enough to save the existing building from closure. Even if the court was extended into the adjoining section of the Crescent occupied by the police (who are due to move to a new station soon), there would still not be enough room for the expanded and modernised Magistrates' Court and the two courtrooms capable of holding Crown Court sessions that the Lord Chancellor's Department is now insisting upon. A new building, PPP-funded, will be built.

This is raising further problems for the Borough Council. Though wary of being saddled with a white elephant, it is equally determined that this civic building should not been turned into a bar or other leisure use once the police and court leave. The leader of the council recently described them with pride as, 'one of the finest civic buildings in the North West if not the country'. The Council's determination to ensure that a redundant civic building has a public future deserves the widest possible support and encouragement. It will be no easy task to develop a sustainable scheme for the building, but nevertheless every effort should be made to preserve Court 1 intact as part of whatever proposal emerges.

Boston Sessions House
Warmgate　　　　　　　　　　　　　　　　　　　　　　　　　　　　　　　Grade II*

Lincolnshire's magistrates built a number of substantial venues for the Quarter Sessions in the nineteenth century, including Spilsby (q.v.), but none more lavish than the near identical Boston and Spalding Sessions Houses, designed by Charles Kirk and completed in 1843. Boston is probably the slightly better preserved, but both are remarkable for the survival and the quality of fittings and furniture.

Boston's creamy ashlar façade is a massive and richly detailed castle, with buttresses and crenellated parapets. Two three storey towers flank the central three bays, which have perpendicular tracery in the tall first floor windows. Above are the Royal Arms and splendid heraldic beasts perched on top of the buttresses. Inside is a large stone-flagged entrance hall, with shallow stairs to the left leading to the back of an enormous courtroom. Its blank ashlar walls create a very sombre atmosphere, but the quality of the woodwork is superb, particularly the richly moulded roof trusses with pierced cusped panels. The only significant alteration has been the addition of a security screen to the dock. The Grand Jury Room is off the courtroom on the side opposite the public entrance. Behind the bench, over the entrance hall and lit by those deep Perpendicular windows, is the Magistrates' Retiring Room, now a second court, with excellent linenfold panelling and marvellous Gothic chairs. Across the sinuous magistrates' stair is a delightful robing room with vaulted ceiling and more original Gothic furniture.

The Sessions House remains in regular use as a Magistrates' Court and the Magistrates' Court Committee takes obvious pride in it. It has recently been reroofed and the stonework repaired. The furniture has been expertly restored and the cells refurbished. With a good claim to be one of the finest Magistrates' Court in the country, it should be fit for many years of judicial service yet.

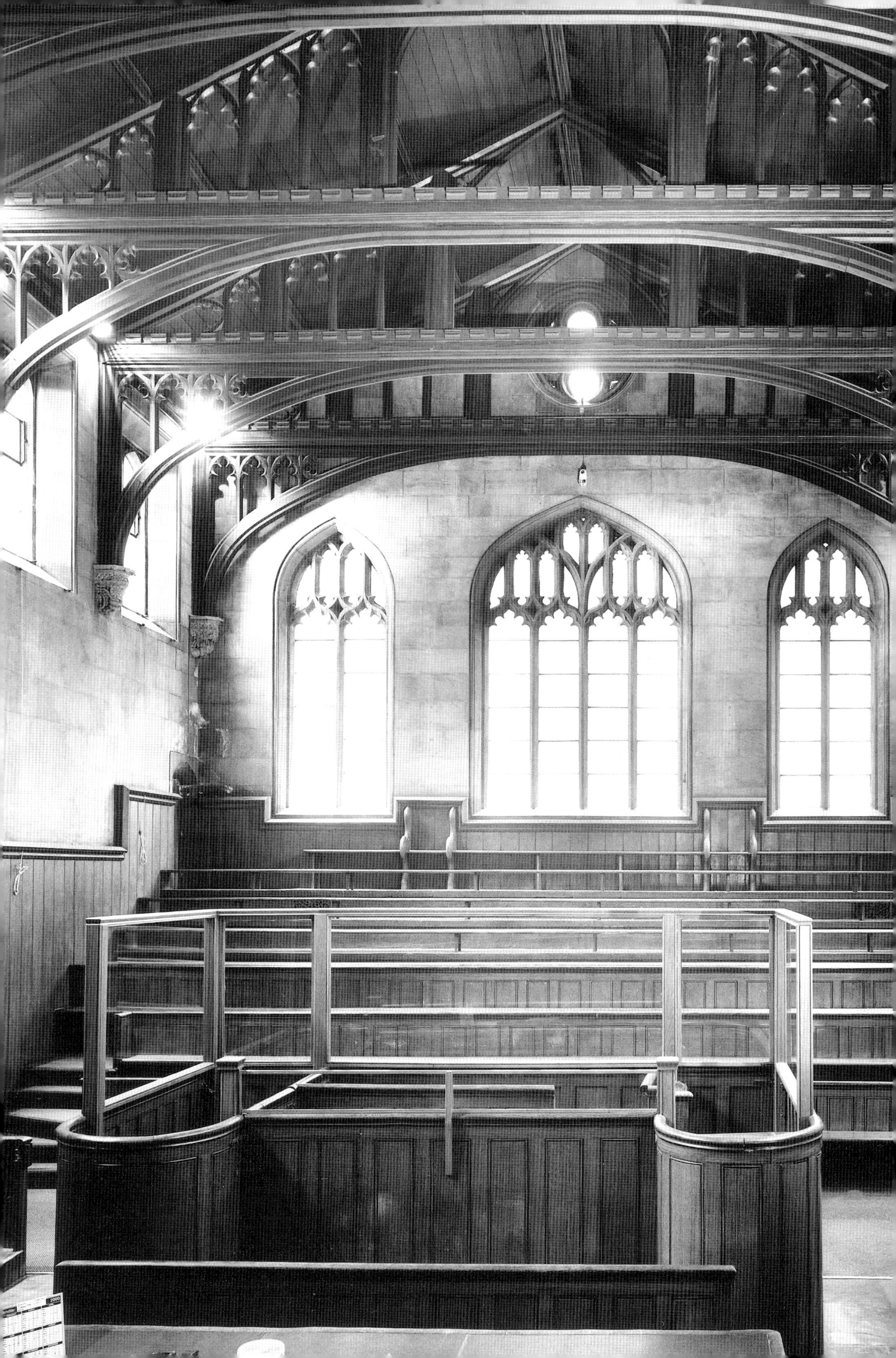

Bridgwater County Court

Queen Street   Grade II*

This is an elegant sandstone building, just off the main shopping street in Bridgwater, begun in 1824 and designed by Richard Carter. The central three bays of five are brought forward under a pediment, supported by engaged Ionic columns. The ground floor is treated as five segmental-headed openings, filled with doors at either end and sashes in the centre. The deep windows create a remarkable area of glass for what is quite a modestly sized façade. The interior has been much knocked about, both when the building first became a court in the later nineteenth century and in subsequent unsympathetic refits. A good open well staircase remains, and the first floor courtroom, set along the front façade, has a panelled segmental barrel roof, but there is not much else of note.

The court closed in 1999. Fortunately, as with other modest county courts, it was quite well suited to alternative uses – because it was already essentially an office building and the courtroom, devoid of interesting fittings, could be reused as a handsome meeting or conference room, or open plan office. It has been acquired by Lyndon Brett Partnership, a firm of chartered surveyors, and refurbished to meet their needs without causing undue concern. The courtroom itself has found an ideal new use, as a meeting and arbitration room, not too distant from its earlier incarnation.

Canterbury Sessions House
Longport   Grade II

The construction of a new Sessions House in Canterbury, to replace a building of 1730 which stood opposite the Castle, was sanctioned by an Act of 1807. The scheme was combined with a project for a new gaol, identified as necessary at the 1804 summer Assizes, and George Byfield, the architect already selected to design the new gaol, reworked his plans for the site in the suburb of Longport to include the courthouse.

The building, completed at a cost of £4,690.15.6 in 1810, was described at the time by Edward Hasted in 'The Canterbury Guide' as, 'equal if not superior to any other in the kingdom; having appropriate places for the transaction of all manner of county business: the courts and offices, etc. belonging to it are fitted up in a style combining elegance with utility'. The Sessions House may not entirely be deserving of Hasted's lavish praise, but the principal elevation is undoubtedly a fine, austere, composition in Portland ashlar, with a central bay emphasized with giant attached Doric columns in antis. As Clare Graham has pointed out, this design was clearly derived from James Gandon's elevations for Nottingham County Hall, as published in Vitruvius Britannicus in 1771 (see Clare Graham's essay). Though most of the ornament was omitted, crossed fasces, and the staff and cap of liberty were incorporated in the central entablature. Coade stone figures of Justice and Mercy (since lost) and iron railings incorporating fasces complete the iconographic schema. These are rare in English courts and Dr. Graham concludes that, 'More clearly than most contemporary buildings, this one communicates the idea of court'.

Internally, the building contained all the facilities then expected of such a building, including a Grand Jury Room on the first floor, connected to a private gallery in the single courtroom. There were two entrances into the courtroom, one for the public seating and one within the bar. Stairs from the dock led down to a tunnel linking it to the adjoining prison. Though this still exists, it is now solidly blocked – because whilst the prison is still very much functioning, the court is not.

Following closure as a Magistrates' Court in 1995, the building was acquired by the neighbouring Christ Church University College, which has spent £3m converting and extending the building to provide lecture theatres, computer and music rooms and staff offices, This work, by Pateman & Coupe, is thoughtful and largely successful: despite the scale of the additions, the Georgian facade is not overwhelmed. Many of the interior spaces have been cleaned up by removing later partitions to restore original room sizes. A 1965 rear extension has been retained but refurbished, its courtroom converted into a lecture theatre. A double height atrium, which functions as the new main entrance, has been created between the 1810 building and the large circular lecture theatre, the drum, which defines the extension's separate identity. The nineteenth century courtroom itself has been turned into a multi-purpose hall / lecture theatre, but regrettably this has required the complete stripping of its fittings. Though they were not original, photographs reveal them to have been substantial Victorian joinery. Now it is impossible to discern what the room's original function was. An acceptable sacrifice? Probably.

Carlisle Assize Courts
The Citadel, Court Square   Grade I

These are some of the grandest, and most unusual, Assize Courts in England. They are unique because the Crown and Nisi Prius courts occupy entirely separate buildings divided by a public road, one in each of two stout red sandstone towers that formed part of the citadel built by Henry VIII to stand guard at the southern entrance to the city. The courts were begun before 1808 by Thomas Telford and John Chisholme, but after the foundations were found to be defective Sir Robert Smirke was brought in to complete them in 1810-12. The structural problems meant that although the eastern tower containing the Nisi Prius Court is substantially original under refacing, the western tower containing the Crown Court had to be rebuilt from ground up. Within each tower is a single semi-circular courtroom wonderfully complete with its original Gothic detailed oak furnishings. Each enclosure has hinged doors like box pews.

A gangway separates the public gallery from the rest of the court, its screen to the back of the dock being nearly six foot tall so that no-one entering or leaving should disturb court proceedings.

The towers are attached to matching Gothic ranges containing ancillary accommodation. These face each other across Court Square, a newly created and handsomely formal civic space, something of a rarity in Georgian towns and cities. The western tower contains the Grand Jury Room, a Petty Sessions Court and a grand stair, with splendid Gothic decoration including plaster vault ceilings. This block also adjoined the county gaol, since demolished, though the cells beneath the Crown Court and the stairs up into the dock remain.

The Nisi Prius court closed with the end of Assizes in 1971, but the Crown Court soldiered on until a new combined court opened in 1992. Redundancy left the County Council, the buildings' owners, with a huge headache. The cost of putting right decades of underinvestment (tackling extensive dry rot, replacing stonework, renewing drainage, replacing heating etc.) is vast. Repairs began on the eastern (Nisi Prius) tower in the 1980s and moved to the western one in the late 1990s, where the restoration of the interiors has also begun at a cost of £1.2m. This is a formidable investment in buildings that will neither support intensive new revenue generating uses, nor attract large numbers of paying visitors (no matter how fascinating they may be). Currently the Crown Court is open for occasional public tours and functions and offices occupy some of the Nisi Prius side, but the long-term future of the buildings remains unresolved. Viability is the key issue for the County, but far greater public access and better interpretation must be at the heart of any scheme.

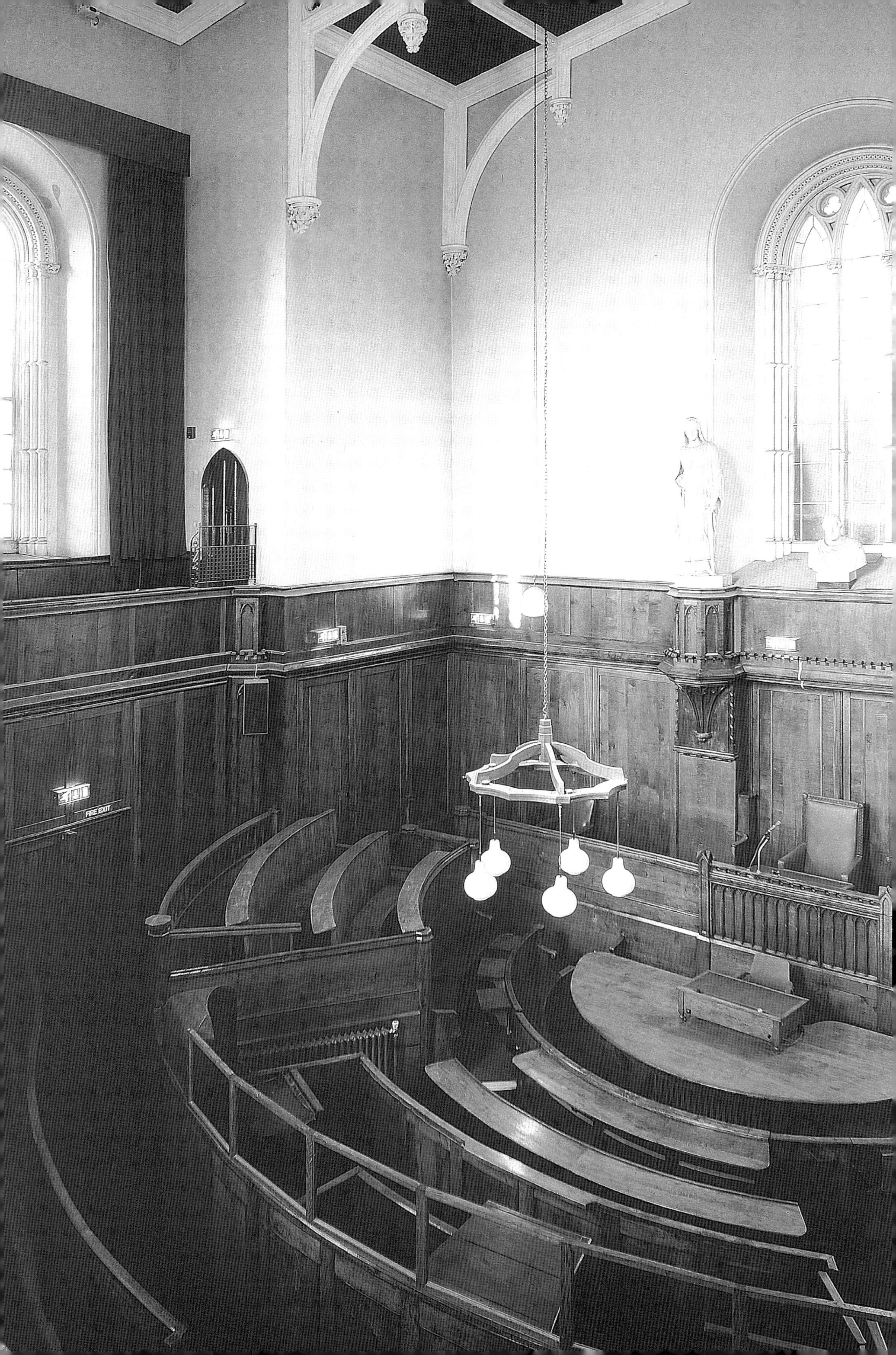
FIRE EXIT

Cheltenham County Court

County Court Road   Grade II*

Assize and Sessions Courts were constructed sporadically over a period of hundreds of years. They were built, not by one central authority, but by individual counties and towns and only two architects designed four or more (Sydney Smirke, seven; Charles Watson, four). Thus, although many share design characteristics and similar plans, they do not form a single coherent group. By contrast, the County Courts erected in the wake of the 1846 County Courts Act were part of a centrally controlled national programme undertaken under the auspices of the Surveyor of the County Courts and many exhibit standard plans and a recognizable house style, an Italianate Palazzo idiom, developed by the first Surveyor, Charles Reeves, and continued by his pupil and successor, Thomas Sorby.

Cheltenham, built between 1869 and 1871 to Sorby's design, is easily the most important surviving example. In many ways it is a typical two-storey exercise with a first floor courtroom. The exterior is by no means the grandest; the cramped and irregular site made that difficult. But inside it is remarkably unaltered, making it the only substantially intact County Court interior to survive. The plan of the courtroom is similar to many others now destroyed. The jury box was originally to the judge's left (allowing the jury to retire to their private room via an adjacent door) and the larger press box to the right. When legislation in 1903 increased juries from five to eight, the two were simply swapped around. Of course there is no dock and no witness stand. Instead, either side of the clerk's desk below the judge's bench, there are the plaintiff and claimant's stands. Beyond, facing the bench, are the curved public benches.

Cheltenham's confined site has now proved to be its downfall. The building requires thorough modernisation, but Lord Chancellor's Department studies have concluded that there is simply not the room available for the expansion that modernisation would require. Therefore it will have to close.

Once that happens, as the one County Court of quality to survive little altered, it ought to be preserved, not reused. English Heritage and the Court Service are fully aware of the significance of the building. With their support and the largesse of the national lottery, should the building in fact be a serious candidate for the National Trust? The Trust doesn't own a courthouse, but acquiring one would fit well with its laudable recent policy of conserving representatives of a much broader range of building types than the country houses and cottages with which it is traditionally associated. These have included a workhouse, back-to-back housing, a textile mill and a Victorian suburban villa. Why not a law court? At first sight the most obvious choice might be one of the great Georgian or Victorian Assize Courts or Sessions Houses, but some of the best examples of these have already been preserved and opened to the public (for example, Nottingham, Bodmin and Stafford).

Cheltenham would be a more important acquisition because no other agency may be willing or able to acquire, restore and manage an example of a County Court. They maybe less splendid than the grand homes of the assizes, which witnessed notorious trials and historic events, but one of these overshadowed and rapidly disappearing but nonetheless important elements of our judicial system ought to be preserved for future generations.

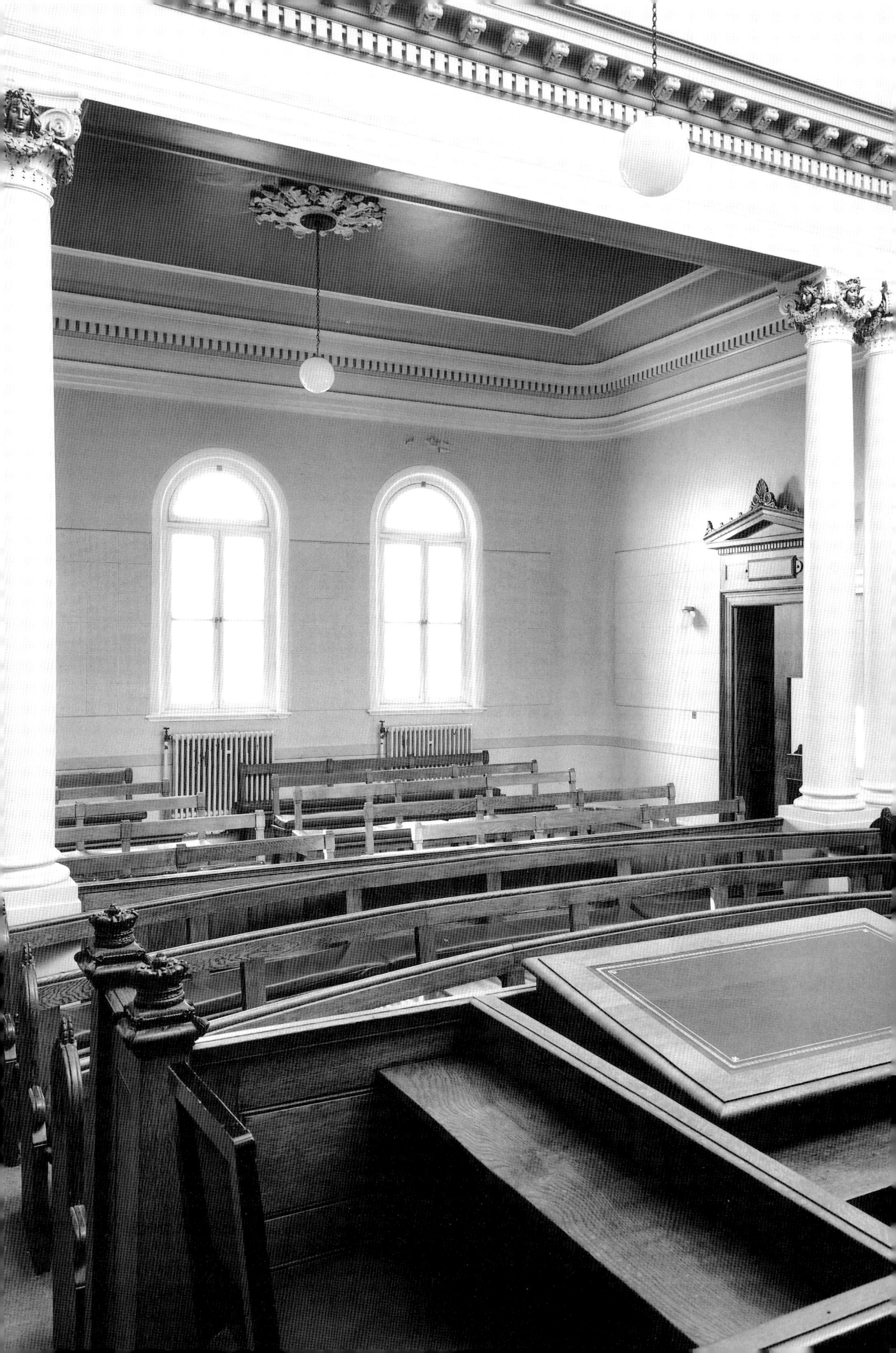

Chester Shire Hall

Chester Castle   Grade I

Chester's Crown Court forms the central element of what Sir Howard Colvin describes as the finest group of Greek Revival buildings in Britain, 'the Neoclassical counterpart to Greenwich Hospital'. The ambition, cohesion and monumentality of this complex of public buildings built at Chester Castle between 1788 and 1822 seem decidedly continental and slightly unlikely in a provincial English city. The architect was Thomas Harrison, born in North Yorkshire, but now indelibly associated with Chester and Lancashire, where he made his reputation as one of England's most accomplished Neoclassical designers.

Entry is via a magnificent propylaeum to a great yard; across this lies the Shire Hall with the prison behind, and flanking are matching pavilions containing barracks and armoury. Behind the mighty Doric portico is the Assizes Court, now Crown Court No. 1, a marvellous semicircular room with an Ionic colonnade and richly coffered ceiling, one of the very finest in the country. The bench is now at the curved end, the reverse of the original seating plan. A Nisi Prius court (now Court No.2) was added in 1875-6 by T M Lockwood. A further courtroom was subsequently created in the Grand Jury Room.

The circulation and ancillary spaces were always cramped, and become totally inadequate with the increased workload of the 1970s and 1980s. To their great credit the Lord Chancellor's Department and the PSA responded not by moving to a new site but by modernizing and expanding the existing building.

This was achieved with welcome sensitivity by constructing a new block behind the Shire Hall in a courtyard enclosed by the old Male Debtor's Prison wing. The architects, the Biggins Sargent Partnership, designed the block as a free-standing building in ashlar, creating an airy atrium by glazing the gap between Harrison's building and this new structure. This became the principal public circulation area. The new block includes an additional courtroom, located on the first floor so as to be top-lit, an example of the recent reaction against the artificially-lit 'bunker' courtrooms of many 1980s courthouses, whose oppressive atmosphere has been heavily criticized. The space below, and more efficient use of the existing structure, enabled the architects to create the extensive ancillary accommodation and segregate circulation routes now expected of a Crown Court.

Since the restored and extended court reopened in 1992 it has served as an excellent model of how historic courts can be imaginatively modernized and expanded. Now all that remains is for Chester to follow the liberating example of Horse Guard's Parade and Somerset House in London and banish the rows of parked cars from Harrison's great courtyard.

Chesterfield Magistrates' Court
West Bars   Grade II

Chesterfield Magistrates' Court is a unique building, the only listed postwar law court. Designed by Professor J S Stanley with R Keenleyside and opened in 1965, it was as one of the first Magistrates' Courts to be built as a stand alone building, entirely separate from the police station. Its design is a liberated response to this new brief, a free-standing three-storey pavilion rising dramatically in the steeply sloping Shentall Memorial Gardens below the 1930s town hall. It is quite unlike any other courthouse in Britain. The plan, two fan-shaped blocks linked at the centre, and the exposed concrete gable frame and deep concrete louvers, resemble that of a university library or lecture theatre complex. This is reinforced by the interiors of the three double-height courtrooms in the larger eastern fan, which are planned as auditoria. The original wooden furnishings remain.

Unfortunately the pavilion plan and parkland setting have now contributed to the building's downfall. Chesterfield has been assessed as requiring seven courtrooms as part of the Derbyshire Magistrates' Court PFI scheme. The only way to provide these and enhanced ancillary facilities and security arrangements at West Bars would be by building a large extension, and the only means of doing this without destroying the fundamental free-standing nature of the 1960s building – burying it – would be prohibitively expensive. Therefore a new complex is being built, leaving West Bars redundant.

So what future for it now? Perhaps council accommodation, with the courtrooms as seminar / conference spaces? Its unique plan probably makes it even more difficult to reuse than most court buildings and any proposal requiring additional space would presumably face the same problems as the PFI scheme. These now are questions for the winning PFI consortium, Derbyshire Courts Ltd (led by Babcock and Brown). Construction of the new court building began in late 2001 and will finish in 2003. Under the terms of the PFI contract, when the replacement building opens the West Bars building will be transferred to the consortium, along with the challenge and responsibility of finding a sympathetic new use. SAVE will await proposals with interest.

West Bars is not the only Magistrates' Court in Chesterfield: there's another on Brimington Road with some handsome Edwardian rooms, including the two courtrooms. It too is listed, and situated close to the station, with court rooms that would make attractive open plan offices, a brighter future for it is easier to envisage.

Chipping Campden Magistrates' Court High Street

Grade II*

The rationalization of rural Magistrates' Courts in the last fifty years has been dramatic. In Gloucestershire once there were twenty petty sessions, each with its own separate court venue. Today magistrates sit in just five towns and all twelve of the courtrooms in the purpose-built combined police stations and petty sessions courts erected by the County in the later nineteenth and early twentieth centuries have closed. Many of these shared a common two and three storey Gothic design, varying principally by the local materials of which they were built. The plans were equally regular: the police on the ground floor, with cells at the rear; the single courtroom and magistrates' room on the first floor. Access to the courtroom for magistrates was from the front (sometimes with a labelled door) and for the public by a covered outside stair around the side. The courtrooms had high ceilings with exposed beams. The fittings were simple and straightforward: a raised bench with a table for the clerk and solicitors in front, flanked by raised dock and witness stand, and faced by public seating beyond a low partition, the symbolic bar.

Now that they have closed, these courthouses have found a wide range of new uses, a good illustration of the inherent flexibility of so many historic buildings. Fairford, Chipping Sodbury and Lydney are still used by the police; Berkeley has been converted into flats; Cirencester is now offices; Newent and Tetbury are museums; Stroud is a Liberal club and offices; Thornbury is a local council office and information centre; Chipping Campden was acquired by a trust established by the town council to turn it into a multi-purpose community building.

Chipping Campden is perhaps the most interesting story and the plan is a good one. The trust leases one room to the police at a peppercorn rent so that they can retain at least a basic office in the village after the police station was withdrawn. Another room is let as a shop and another is now the tourist information centre. The rear ground floor room will become a pre-school playgroup and the courtroom is to be made available for meetings, talks and seminars, both for community groups and clubs, and for commercial hire.

Now you would have thought this exemplary scheme, run by the community for the community, would have been warmly welcomed by the County Council, which owned the building and which wanted to dispose of it after the departure of the magistrates and the police. The town was even offering to provide free accommodation for the police. Yet it took a year of vigorous lobbying by the locals, their MP, County Councillors and the Gloucestershire Rural Community Council before the County finally agreed to sell the building at a price the town could afford, which was understandably less than the open market value that the Treasury requires local authorities to obtain when disposing of surplus property

This highlights the absurdity of the rules governing council disposals, which here almost prevented the transfer of a public asset, built at public expense, to a local community which wanted to use it as a public resource. A change of Government policy is required – an unambiguous commitment to offering all redundant publicly-owned buildings first for community or charitable use at below market rate. Only if no viable proposals from this sector come forward should they then be put on the open market.

Derby Shire Hall
St Mary's Gate   Grade 1

This has been a cause célèbre amongst historic law courts for fifteen years.

The Shire Hall, erected at the end of the Commonwealth, in 1659, is the earliest surviving building purpose-built to house the assizes. Originally the Crown and Nisi Prius courts sat at either end of the single open hall, with their own separate entrances, still visible today in the St Mary's Gate façade today. They may have been separated by a central section containing Grand and Petty Jury Rooms above one another, though the evidence is inconclusive. The heavy mannerist façade stands in a courtyard 'once graced with an avenue of trees' (Hutton's History of Derby, 1817), and gardens were laid out to the rear.

As the assizes were only in Derby for a few days a year, the Hall was also built to stage many other public events, such as plays, dances and concerts. This tradition continued right up until the late 1990s, when the High Sheriff hosted his final reception there. However, as alternative venues opened in the town and as the business of justice became more complex and the demands placed upon the assizes grew with Derby's population, the court facilities expanded and most of the other uses were forced out. In 1798 a County Hotel was built along the west side of the courtyard to house advocates and witnesses. In 1809 the Marlborough Head public house was demolished to make way for new Judges Lodgings on the east side. Twenty years later the complex was extensively modified when Matthew Habershon (architect of Derby's Town Hall) added two new courtrooms to the rear of the seventeenth century building (replacing a Grand Jury room of the 1770s by Joseph Pickford) and a new service block behind these. At the same time the Shire Hall was gutted of its court fittings to become a grand entrance hall, with a balcony along the rear wall providing access to the public galleries of the new courts. Finally, in 1881-2, after complaints about their build quality and the inadequacy of the ventilation, Habershon's courtrooms were rebuilt by the County Surveyor, J Somes Story.

In this state more or less (though the County Hotel became successively the County Library and a police station) the complex continued as a law court until 1989, when it fell victim to the Lord Chancellor's court building programme and was replaced by a new courts complex elsewhere in the town. Most of it has been unused since then and parts, notably the County Hotel, have fallen into the most deplorable condition – riven by rot and structural failure.

In 1995 there was local and national uproar at plans to refurbish the complex as the new Magistrates' Court complex for Derby. These proposed demolishing the County Hotel and replacing it with an office block and demolishing the 1828 service accommodation (including the Grand Jury Room) and replacing it with new courtrooms. The proposal was abandoned, but the need for a new Magistrates' Court remained and with a change in government policy this requirement re-emerged as a PFI scheme for the whole of Derbyshire, one of the first for the Lord Chancellor's Department.

This put the future of the Shire Hall in renewed doubt. PFI

bidders did not want to be saddled with a shoe-horning a dozen new courtrooms into a cramped site currently filled by a Grade I listed building and other listed buildings in appalling condition. The existing Magistrates' Court on Derwent Street offered a less risky, and less expensive alternative. However, the Court Service, the County Council, the City Council and English Heritage all recognized that if the Magistrates' Court did not come to St Mary's Gate it would be extremely difficult, and costly, to find any other suitable use for the Hall. With a welcome sense of responsibility the Court Service therefore altered the terms of the PFI competition to make reconsideration of the Shire Hall a mandatory variant of the contract and told bidders that the Lord Chancellor's Department would increase grants to bridge the extra cost of this option, though, if the site could not provide the facilities required by PFI, they would not be forced to include it in their bid. The scheme now became a 'pathfinder project' to identify means of integrating the requirements of listed buildings into the PFI process (Chesterfield also has a problematic listed Magistrates' Court which forms part of the PFI scheme, q.v.).

The upshot of this was the eventual selection of Derbyshire Courts Ltd (led by Babcock and Brown) as the preferred bidder consortium, with a commitment to develop the Shire Hall site. Online Architects produced a scheme in which the biggest improvement over the 1995 proposal is certainly the retention and restoration of the County Hall to provide support facilities for the complex. But as with the 1995 scheme, everything behind the 1882 courtrooms is cleared away to make way for new courtrooms – a further ten in total.

In principle this is a necessary sacrifice, because it is this that makes it possible to return the building to use as a law court. In plan the solution is admirably clear and simple: the Shire Hall itself is restored as an entrance hall to the two nineteenth century courtrooms and the ten new ones which are laid out on one level off a single U-shaped lobby wrapped around the listed building. None of this will be visible from the restored courtyard.

Sadly, however, the scheme is let down badly by the poor quality of its architecture. The elevations of the new wing are leaden-handed, graceless, mean and cheap-looking. For this, there really can be no excuse: this is an important civic building and a Grade I listed structure. The junction between the new block and the County Hotel is particularly grating. The lobby linking the new courts around the exterior of the listed building is also a weak point; hemmed in between the walls of the Shire Hall and the wrap-around extension, housing the new courtrooms, it must be top-lit, but this seems to have been done in the most measly way. This will be the most important public space in the building, filled with nervous defendants and family and friends, harassed solicitors and press and police. More thought and attention should have been given to creating a generous, calming environment.

Derby is an important project, because PFI/PPP is the future, for better or worse, of capital investment in the public services. The Lord Chancellor's Department and Court Service need to be able to demonstrate through Derby to future bidders and clients that complex listed buildings and PPP can work. Clearly hard-nosed private sector consortia have to be won over, but more importantly so too do Magistrates' Courts Committees. They, not the Court Service, are the clients, and they may equally view run down historic courts as the problem, not the solution. Faced with a Victorian building that has been starved of investment for decades, it is unrealistic to expect them to fight for its refurbishment against the shiny new complex proposed by the bidding consortia.

The future of Waterhouse's Bedford Shire Hall and the incomparable Victoria Courts in Birmingham amongst others are dependent on the outcome of the PPP process. Derby is crucial to demonstrating that the nature and design of PPP does not have to mean the closure of listed court buildings. On completion the consortium will be able to assess the real financial cost, and users the benefits, of retaining a historic building on a city centre site.

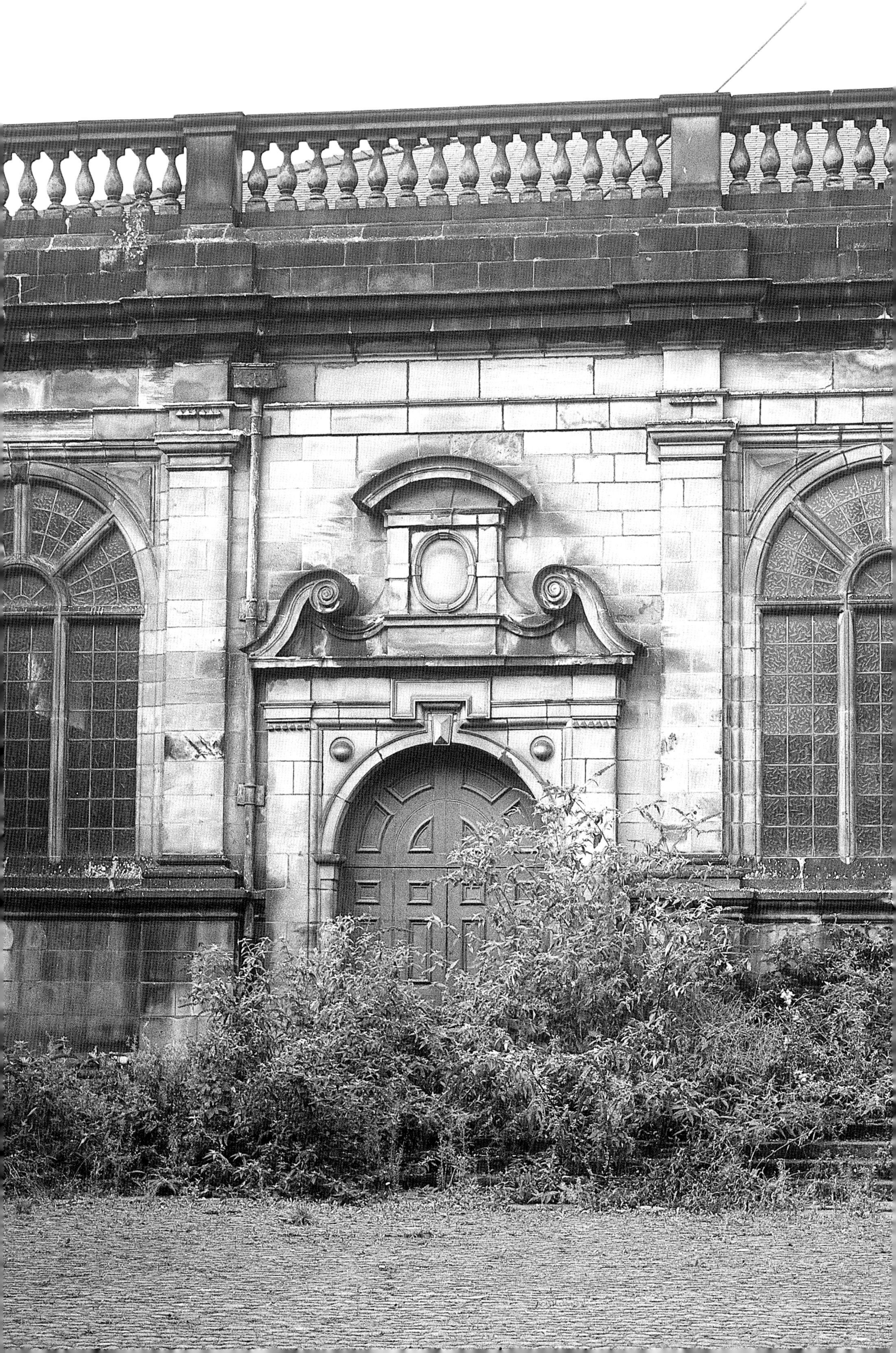

Devizes Assize Courts
Northgate Street   Grade II*

A disgraceful state of affairs. T H Wyatt's dignified building, dominated by a monumental Ionic portico, closed as a Magistrates' Court in 1982 (the Crown Court left in 1975) following an outbreak of dry rot. Though this was eradicated, the magistrates never returned and since 1986 they have sat in portacabins. Since closure and disposal, five development plans have come and gone and permission has been granted for office use, a health club, restaurant, residential use and a night club, and yet all that has changed in that time is that the building has been gutted and now stands derelict.

Though Salisbury was always the principal assize town of Wiltshire, in 1834 Devizes successfully petitioned the Privy Council for the right to hold one of the Assizes and raised a subscription to erect a suitably dignified building to house them. Wyatt was commissioned and he produced a design, as Dr Clare Graham has pointed out, that was clearly influenced by Robert Smirke's neo-classical courts in places like Maidstone, Gloucester and Hereford. A double-height coffered entrance hall behind the portico gives access to the two courts beyond and, flanking it, rooms that probably functioned as the Grand Jury Room and a room for counsel. The two courts, with public galleries against their back walls, are lit by first floor windows on the other three sides. At closure the courtrooms retained much original joinery. This was of the simple box pew type, on a rectangular, not curved, plan. But bench, dock, jury box etc. have all now been removed, leaving only a rubble-strewn floor.

It is claimed that Devizes suffers from two major problems which have prevented commercial reuse to date. The first is common to most court buildings, the legacy of course of their highly specialized planning, in that most of the volume is taken up by the major double-height public rooms, particularly the courtrooms, which cannot be subdivided, whilst the rest is inconveniently planned cellular accommodation. Thus, it is seen as an unproductive, unviable building. These problems are not insurmountable, particularly if, now that they have been stripped, a mezzanine floor was inserted in one of the courtrooms. Second, it has been argued that the court house has inadequate parking (it stands on a busy main road), though this is also disputed: fifteen cars could be parked on site, and the nearest car park is only four minutes walk away.

The current owners have secured permission to convert the building to residential use, although because this will mean the subdivision of both courtrooms, this is far from an ideal solution. There also some concerns about the commercial viability of the scheme. Though the owners have yet to act on the permission, they have at least, when faced with an Urgent Works Notice served by the North Kennet Council, undertaken more extensive repairs than was strictly required.

At the same time, major investment is being planned in the regeneration of the adjacent district along the Kennet and Avon Canal known as North Gate. The courthouse is the most distinguished building in the area. Therefore, it ought to be a priority for action and investment by the agencies overseeing the project. This could be just the breakthrough needed to finally get a scheme for the building off the ground.

If all else fails, North Kennet must take the lead by brokering a solution with a buildings preservation trust or sympathetic developer and serve a repairs notice and, if necessary, a compulsory purchase order. It is over twenty years since the court closed. At some stage the council must be prepared to use its powers to intervene and prevent the building from standing empty for another twenty.

Downham Market County Court
London Road Grade II

Another very typical Charles Reeves product, built in 1860-1. Here his favoured single storey plan is adapted to fit a narrow, crooked site. Only a modest three bay façade faces the road, but the three bay return is embellished too, as it is exposed across the entrance to the police station at the rear. The double height courtroom behind, with large arched windows, could not be set perpendicular to the front block either; the junction is a little top-lit lobby that served as the public entrance.

The court closed in 1991 and stood empty for seven years – proposals for a restaurant came and went – before it was acquired by the present owners for conversion into a house. As at Maldon (q.v.) the appeal was its unconventional possibilities and the drama of the huge courtroom. In addition, the small attached police house provided more conventional domestic spaces. Throughout, good use has been made of the original fittings: the corridor down the side of the courtroom retains its long wooden benches on which generations waited for cases to be heard; in the courtroom itself the dais and the jury box remain in place and deal panelling is set back against the wall as wainscoting, and along the front of the balcony that has been introduced at the back of the court. The new owners have made the most of the possibilities thrown up by their unusual building to create a unique and much loved home.

85

Durham Crown Court
Old Elvet   Grade II*

For many centuries justice was administered in Durham not by the Crown, but by the mighty Prince Bishops, who could hang a man, raise an army and mint currency. But from the Middle Ages these peculiar powers were curtailed by successive monarchs and the assize judges had been travelling up from London for hundreds of years before the historic site of the assizes, under the shadow of the Cathedral, was abandoned for the present building in 1811.

The building had an awkward birth. In 1808 Justice Sir George Wood recommended that new courts and gaol be erected, and with the aid of a gift of £2,000 from the Bishop of Durham the foundation stone for a new court across the river, off Old Elvet, was laid on 31st July 1809. The original architect was Francis Dandys, commissioned on the strength of his Gloucester Gaol. He, though, was soon dismissed for fraud (the Durham Magistrates successfully sued him for a staggering £20,000) and the fabric was taken down, but his replacement, George Monneypenny, was soon himself replaced, this time by Ignatius Bonomi, a Durham man and county surveyor of bridges. In what were presumably considered his safe pair of hands the new building was completed in time for the 1811 summer Assizes, paid for by a penny on the local rates. Between 1815 and 1819 Bonomi completed the adjoining Gaol, which still stands as a high security prison.

It is Bonomi's facade we see today, a hunched sandstone edifice, enriched by a giant portico of attached Tuscan columns. Flanking the court on either side are rusticated arches, one giving access to the prison, and beyond them attached villas. About the only vertical accent is the small cupola above the pediment. The four square patches above the main door are reputed to fill the sockets into which scaffolding was inserted for the first hanging at the new courthouse, that of John Crieg, executed on August 17 1816 for the murder of Elizabeth Stonehouse.

Bonomi's interior, however, was completely remodelled by the County Architect W Crozier, in c. 1870. The Crown and Nisi Prius courts flank a spacious toplit double-height hall in which an imperial staircase rises through arches carried on superimposed Tuscan columns to the Grand Jury Room and a balcony that extends around all sides of the hall. Off this, above the entrance and supported by more Tuscan columns, is the magistrates' meeting room. The courts themselves retain almost all of their original decoration and joinery, excellent examples of late Victorian court design. There are galleries on three sides and a full-width canopy extends over the bench. The panelling behind the bench picks up the motif of a blank colonnade carried around the upper walls of the courts and the balcony of the hall; the varnished oak panelling in the well of the court and on the balcony fronts is enlivened by coupled Tuscan pilasters.

Throughout the building a wealth of original detail survives, enhanced by a recent refurbishment (which includes a carpet in the hall copying the pattern of the lost 1870s tiled floor): lamp standards in the courtrooms, acanthus scroll embellishment above the Judge's chair, and bishop's orb finials to the cast iron balustrades found everywhere.

By all accounts Durham is a successful court; for many years it has been the most efficient court on the North East circuit, despite its age. It is an excellent example of how historic courts can meet modern judicial requirements without requiring substantial alteration and the destruction of fittings and character and it will hopefully continue to fulfil the function for which it was designed for many years to come. However, Bonomi's infamous jail may not be so fortunate. HM Prison Service has floated proposals to close the remaining inner city Victorian jails. Eventually, students, not criminals, could be living in the cell blocks behind the court.

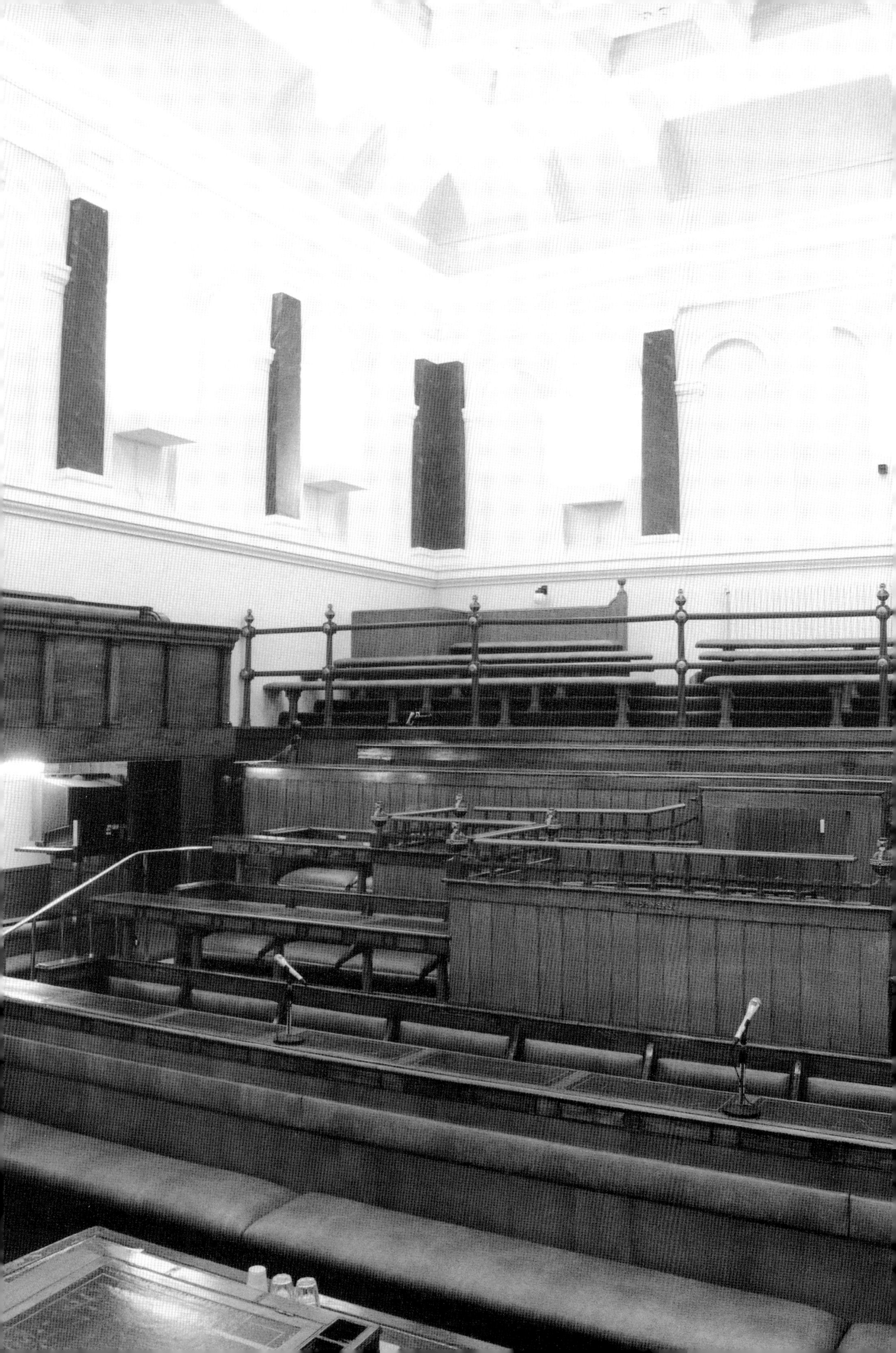

Ely Shire Hall
Lynn Road   Grade II*

Many new courts were erected between the late eighteenth and mid-nineteenth centuries – specialist buildings built to cope with the demands put on the judicial system by the rapidly expanding and urbanising population. Many of these buildings are dignified, though hardly magnificent, relying on scale, massing and an imposing portico rather than expensive materials and lavish architectural details. As Dr Clare Graham notes, English architectural practice at this time did not appear much interested in theoretical debate about type and style. The emphasis was on convenience and economy. Thus, for example Joseph Gwilt, writing in 1842 in his Encyclopaedia of Architecture, said of courthouses that architects should concentrate on avoiding the common error of paying too much attention to well-paid judges and barristers, while leaving everyone else 'pent up and cramped like the poor sheep at Smithfield'.

Ely Shire Hall is typical of this type of building. A simple yellow brick structure, given stature in a small country town by a Greek Doric portico of Roman plaster (no money for real stone), it was erected at a cost of £6,000 by 1822 to the designs of Charles Humfrey. Decoration is restricted to the county arms in the pediment, a common emblem in these county-funded civic projects. At the rear was later constructed an octagonal house of correction; only a fragment remains.

Happily Ely is still operating as a Magistrates' Court and has recently been refurbished. The single court room is a delight, retaining apparently all its tall enclosed Georgian joinery. Magistrates, in front of a curved rear wall, sit behind a delicate curved balustrade; the jury benches are to one side and the witness box to the other. Completing the intimate arena across the red baize table is the dock. No reinforced glass here, only a turned baluster screen. Behind and above is the public gallery and to one side the Grand Jury's balcony, set behind another balustrade in an alcove. Outside a good staircase connects the various floors, with simple double doors at the foot, and above charming oval lights. The original cells (though with new doors) are still used.

The story of the building is told on panels in the public waiting room, along with the display of two of the original cell doors. Unfortunately, though, here Humfrey fell foul of the crime later identified by Joseph Gwilt – the public spaces are very, very cramped.

Exeter Sessions House
Rougemont Castle   Grade II*

Exeter Combined Crown and County Courts currently sit in the Grade II* Sessions House of 1774, which stands within the inner bailey of the Norman Rougemont Castle. It replaced an earlier Sessions House of 1607, and was designed by local architects James Stowey and Philip Jones, with advice from James Wyatt. In all probability, as built, the two courts flanking the central entrance hall were probably open to it, as they were at James Gandon's Nottingham Shire Hall (q.v.). However, little now remains of the eighteenth century interior after successive brutal refurbishments, and the restrained ashlar facade with open arcade and pediment to the centre three bays is now flanked by a simple twentieth century extension on one side and a rather grander Italianate return wing of 1905 in matching Bath stone on the other.

Now, after hundreds of years of continuous court use on this site the future of the building is in grave doubt. The existing facilities are wholly inadequate and need to substantially expanded. More courtrooms are needed and, as usual, far more ancillary accommodation. This is to be provided by a PFI scheme.

The most important factor in determining whether or not historic courts can be upgraded is normally the availability of adjoining land upon which an extension can be erected to house new facilities. If the brick extension to the left was demolished and replaced by a new wing balancing the 1905 wing to the right (which would revive the symmetry of the classical composition, albeit around a new three-sided courtyard) enough new space could be created to provide all the facilities required for criminal business, but according to Courts Service studies, not for the civil business as well. The only way to provide enough space for all the new facilities required would be to erect a much larger extension that would dwarf the Sessions House. (This is actually close to proposals from the 1930s, which envisioned a four-storey block on this site. War intervened before the project could be fully realized, though part of the 1770s building was remodelled.)

The new court building is therefore to be erected in the legal quarter of the City. It will be one of the first fruits of the recent attempt by the Court Service, the Lord Chancellor's Department and CABE to embed principles of design quality into the PFI process. The chosen design is by the Baptie Group. We await its completion with interest.

As to the Sessions House, the intentions now is to sell it to the City Council, which plans to use it to host as yet unspecified cultural and heritage function. At least one really positive civic benefit of the court's move will be the lifting of security restrictions around the building, allowing more of the castle walls to be opened up to the public. Nevertheless something tangible will be lost when a history for dispensing justice on this site stretching back nearly a thousand years comes to a close.

Grantham Magistrates' Court
London Road   unlisted

As with so many others, Grantham Magistrates' Court is an unambitious but pleasant building. Like a number of other law courts it was built for other purposes, in this case as a private house in 1810. Ten years later it was purchased by the JPs for petty sessions sittings and it subsequently became the police station too. In the last quarter of the nineteenth century a purpose-built courtroom was added in a big extension to the rear.

The police moved out in 1955 and the magistrates finally followed in 1995. After four years standing empty it was bought by a local businessman who was attracted by the sturdy Ancaster stone exterior and the fascinating fixtures and fittings inside – magistrates' bench, cells and all. More importantly he saw the building's potential as both a home and an office and has set about converting it with great enthusiasm, creating light and spacious rooms. The courtroom itself, in which the false ceiling has been torn out, has been turned to an impressive and flexible seminar room, used for staff meetings and training.

Nothing about Grantham Magistrates' Court is spectacular, not its appearance nor its most recent conversion, but it demonstrates once more that historic buildings are flexible and sustainable resources capable of absorbing numerous changes. No expensive alterations or interventions were needed to give a building that has already had two different uses a successful new one.

Simply by exploiting the inherent virtues of the building – working with an open mind with the grain of the structure – the current owner has created an attractive home and pleasant office.

Huntingdon Town Hall

Market Hill

Grade II*

© Clare Campbell

The Town Hall is a pretty pedimented red brick building, topped by a cupola, built in 1745. Its probable architect and builder was Benjamin Timbrell, who, though he did on occasion act as an architect, was much better known in London as the most prominent speculative builder of his time (he was also partially responsible for the structural woodwork of St Martin-in-the-Fields). His building was completed in typical town hall fashion with an open arcaded ground floor used by the market, but this was largely filled in by Samuel Pepys Cockerell in 1817-18.

The building was the venue for the assizes, as Huntingdon was the county town of Huntingdonshire, and it remains a Magistrates' Court today. There are two small courtrooms, both fitted with balconies and grained pine furnishings. Some of these are undoubtedly eighteenth century, but both rooms have clearly been reordered over the centuries, creating the present, pleasingly chaotic and intimate spaces.

But such picturesque attractions are of little benefit to the operation of the court. Facilities are obviously inadequate, but as the building is an island, the opportunities for expansion are clearly limited. Only by filling in the last remaining section of the colonnade can additional accommodation be created, and this is exactly what is now proposed. Although the loss of the last section of the arcade is a matter of some regret, the departure of the court would be infinitely more undesirable. So, subject to the implementation of a sympathetic and thoughtful scheme, this proposal is to be welcomed as a positive attempt to retain the use for which the building was designed.

Kendal Town Hall
Highgate   Grade II

Courts and councils have been bed fellows for centuries. Until the late nineteenth century justices of the peace were the county administration, responsible for roads, bridges etc. When County Councils were created in 1889, unsurprisingly, many were located in, or adjacent to, the court houses or county halls from which the JPs had previously administered the county. The Nisi Prius court sometimes doubled as the new county's council chamber; since this did not have a dock, it was not too dissimilar in design to a purpose built council chamber. Oxford Castle is an example of this. Within towns up and down the country, borough and petty session courts often shared a chamber with the town or borough councils in the guild, market or town hall, such as at Totnes, Chippenham, Rye or Abingdon. Dozens of these survived as Magistrates' Courts until the last few decades. Their simple and flexible design and the fact that the court was always only one use amongst many has meant that the end of court sittings has rarely brought about substantial changes to the room or its function.

In the nineteenth century grander town halls appeared in many towns to oversee the more complex and extensive administration of larger and wealthier areas. These were often equipped with specialist courtrooms, retiring and waiting rooms and separate court access from the street. However, by the end of the twentieth century many of these courts, so sophisticated for their time, were inadequate, unable to cope with the rising level of court business and the amount of support accommodation now required, and consequently many have been made redundant by new, separate Magistrates' Courts buildings. Sometimes this has resulted in pressure to convert the redundant spaces to council use. In other circumstances, where there is not such pressure on space, or the budget for more accommodation, courtrooms have been allowed to quietly gather dust; because their presence adds nothing to the overall maintenance costs of the building, doing nothing costs nothing.

Kendal Town Hall is an example of this which could stand for many others in the country. The earliest part of the building is the 1825 White Hall Assembly Rooms by Francis Webster, which stretches fourteen bays back from Highgate (the chief thoroughfare of the town) down Lowther Street. This was converted into a Town

Hall in 1859 and extended in 1893 by Stephen Shaw, who topped it with a splendid Baroque tower. The courtroom has independent access from Lowther Street and rich dark panelling. A public gallery is carried on cast-iron pillars. Natural light comes from windows on two sides and a large skylight.

In 1991 the Magistrates' Court left the by then totally inadequate facilities for new premises. The fittings in the courtroom were largely intact, the well dominated by the dock with its brass rail. It was decided to convert the room into a new council chamber and this was achieved quite successfully, though it was necessary to sacrifice all fittings in the well, including the dock, to create the ranks of councillors' seating. This was a variation of the normal compromise in which the bench and a few other fittings, possibly the witness and jury boxes, or the public seating is retained and the rest of the space reused. It can only ever be partially successful in retaining the character of the room, just as when pews are removed from a church and replaced by loose seating or tables. As with churches the balance to be struck is one between the need to find a productive new use and the relative significance of the building and its fittings. In the case of Kendal, the sacrifice may not have been necessary after all: the council chamber is being relocated again because of complaints about the acoustics and other arrangements.

Keswick Magistrates' Court
Bank Street   Grade II

The exterior of this combined police station and police court is pleasant, but modest, and beautifully built, of slate with red sandstone detailing. It was designed by the County Architects Department and opened by the Chief Constable in 1902. The police station is no more or no less than an attractive house of the date. Attached to one end is the courtroom, the gable end facing the street with a large Venetian window. Down its flank is the handsome Gibbsian surround to the Magistrates' entrance. The best external feature of all is the public entrance, a pedimented porch set in the re-entrant angle, supported at one end by a free standing Doric column and the other by a console. The inner porch retains original doors and coloured leaded lights. The vestibule has been mucked about unsympathetically, but turning left through a door the visitor finds themselves at the back of a charming and little altered courtroom.

Under the barrel-vaulted roof a couple of simple benches are provided for the public. The bar is just that, supported by turned balusters. Three serried benches face a table and across that, the clerk's desk and the magistrates' bench raised behind that. To the left is the canopied witness stand, with lovely inscribed scrolly brackets; to the right is the dock, fenced with brass rails. The panelling is simple but elegant. On top of the bench are two brass lamp stands. The final surprise is found through the door behind the dock, a row of pristine cells retaining their beautiful iron doors and door furniture.

Throughout the building, the quality of the workmanship and materials stands out. In its modest way, this is as appealing a small rural courtroom as you could see.

Keswick has recently, and rightly, been listed Grade II. It was closed in 2000 and has stood empty since then whilst its future has been considered. One idea that has been floated is to convert it into a one-stop-shop for the district and county councils.

Knutsford Sessions House

Toft Road   Grade II*

A distinguished neo-classical design by George Moneypenny, built 1817-9. In plan it follows the norm for single courtroom sessions houses of the era, with lower wings flanking the central body of the building containing the courtroom aligned centrally and perpendicularly to the entrance front. Compare it with Beverley Sessions House for example (q.v.). Rather than the standard pedimented portico, however, Moneypenny designed an Ionic tetrastyle portico in antis, surmounted by a pedimented lantern. The facades of the flanking wings are blank with the exception of a door set in a spectacularly oversized Vanbrughian surround, an early example of Baroque revival. Perhaps they were inspired by the sublime architecture of George Dance's Newgate Prison, because Moneypenny specialized in designing prisons and the Knutsford commission included a gaol behind (a supermarket now stands on the site). The front is ashlar. The rear is brick with stone dressings, the form of the courtroom clearly expressed by its polygonal end and large windows.

Inside it is a big room, with good ceiling plasterwork and much nineteenth century woodwork, no doubt altered and adapted over the years. The bench in front of the canted end wall follows its curve; at the opposite end the public gallery encloses the porch into which the central door opens. This is no longer used and access is via the right-hand door, which also gives access to the later nineteenth century top-lit Court 2, which is attached to the end of the flanking wing behind the high gaol wall.

The Crown Court still sits at Knutsford, but years of inadequate expenditure on upkeep had left the building in need of substantial repairs, particularly to the roof, and requiring thorough modernization. The Lord Chancellor's Department announced that it was to close. Given the decades-old march towards fewer, larger combined facilities this was not a surprise. An alliance of local citizens, lawyers, magistrates, journalists, councillors and MPs refused to accept the plan and campaigned vigorously and tirelessly until, recently, the Lord Chancellor's Department announced that it was overturning its earlier decision and would instead invest £1.5m of its own on repairing and upgrading the Sessions House to enable it to continue to operate as a Crown Court. In certain circumstances, it then appears, large-scale capital works can still be funded direct from the public purse and not through PFI.

This is a wonderful triumph for local will and determination and should be a source of inspiration for other communities fighting court closures.

Lancaster Shire Hall

Lancaster Castle Grade I

Lancaster Castle, today, is unique. Many provincial castles, remaining in Crown hands, developed over the centuries as the seats of county justice and government, home to the gaol and the assizes. Lincoln, York, Oxford and Chester were all like this. At three of these the court (now the Crown Court) still sits, but only at Lancaster is the gaol still operational too. And what a remarkable place it is. First the position, on a hill overlooking Lancaster, with a splendid many-towered and castellated silhouette. The Norman keep survives, rebuilt higher up, at the centre; so too do medieval walls and towers, including the fifteenth century gatehouse. Most of what now gives the place its character, though, is much later, designed by Thomas Harrison and finished by J M Gandy from 1788 to 1821 in, naturally, Gothic. First came the Gaoler's House, then the female's felon's prison, either side of the gatehouse. In 1793 the prison for male felons – now HM Prison Lancaster – was built with four-storey castellated tower cell blocks radiating north of the keep; a debtor's prison over an arcade south of the keep went up at the same time. Finally, a female penitentiary was added in 1818-21. All this survives. If and when the prison closes (which it undoubtedly will do at some stage) it should be opened to the public as what would be one of the most fascinating historical sites in the country.

Meanwhile Harrison designed a Shire Hall and Crown Court north of the keep, to replace the medieval great hall, which was mostly completed in 1798. From the North these form a symmetrical group, with the seven-sided Shire Hall projecting between two towers, one medieval, one Harrison. Before the project was complete Harrison was asked to resign, the JPs complaining that he had not paid enough attention to the project, and in 1802 the reigns were picked up by Joseph Gandy, who designed the window tracery and the courtroom furnishings. The Crown Court is a relatively straightforward rectangular room, but the Shire Hall is spectacular. It is a seven-sided semi-circle. Over the bench against the long wall rises a broad arch. From this radiates a vault to the arches of an ambulatory around the outside wall. All this with blind tracery plasterwork. The superb Gothic woodwork by Gillows, in a semi-circle facing the bench, is excellent. Gandy's pinnacled plaster canopy over the bench is wonderful. All this makes an interesting comparison with Harrison's Shire Hall at Chester Castle (q.v.), built at the same time. The semi-circular plan of the courtroom is similar, but Chester was Neoclassical throughout, presumably because so little of the medieval castle there survived.

The future of the court was threatened in the 1990s when the a Combined Court Centre at Preston opened, but a forceful local campaign, led by the Law Society and backed the City Council, earned Lancaster a reprieve as a level three court; cases posing a more serious security risk would go to Preston. The campaigners argued that Lancaster was a major population centre serving a wide rural area and that relocation to Preston would cause clients additional travel costs and inconvenience. It was important to local justice that the Crown Court stayed in the city. Nevertheless, because it has only limited facilities and only two courtrooms, the long-term future of the Shire Hall must still be considered uncertain.

97

Leicester Castle

Castle Yard Grade I; Scheduled Ancient Monument

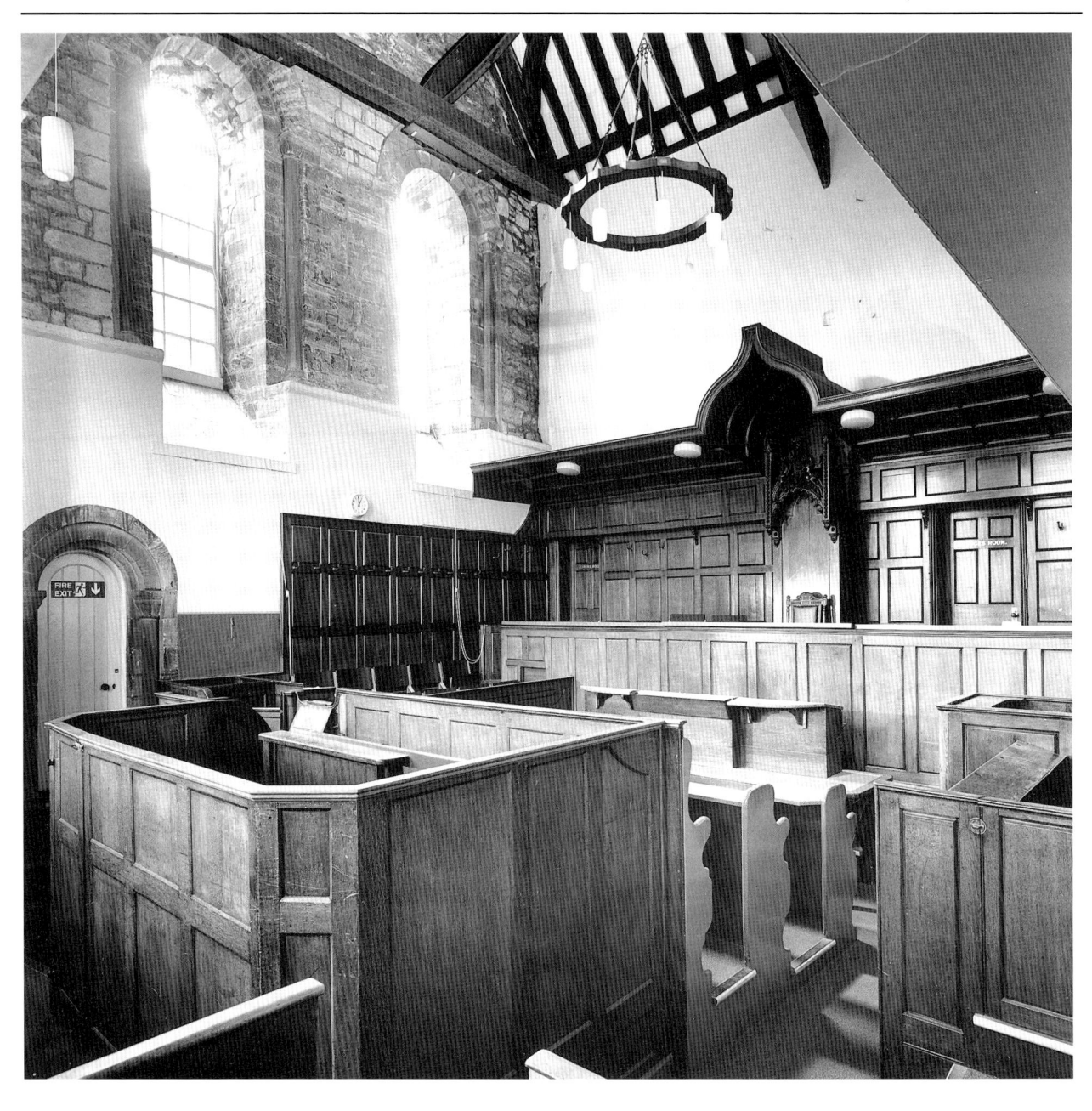

The Great Hall at Leicester Castle is one of the most remarkable examples of the extraordinary continuity of many seats of justice. By the time that the Crown Court moved out in 1980, the building had housed courts of one kind or another for over 800 years. The assizes themselves first sat here in 1273.

This is not immediately apparent. Approached through the picturesque gatehouse (medieval, timber framed with sixteenth-eighteenth century additions; still the judges' lodgings) and set within the intimate grassed yard of the old inner bailey of Leicester Castle, is a large red brick facade, of c.1695. The centrepiece is unsatisfactory, a wide open pediment capping an undersized Venetian window above an underwhelming door. These elements are too small and incoherent for the large windows that flank them (though, in defence of the seventeenth century architect, the pediment was closed until the Venetian window was inserted in the eighteenth century. The door too is later). Rising behind is a high, steeply pitched roof, the first indication that all is not what it seems.

Inside a plain lobby leads left and right to the normal arrangement of Nisi Prius and Crown Courts, both retaining all their 1821 fittings – benches, panelling etc. These are impressive for their completeness, not outstanding quality, for they lack either delicacy or grandeur. Galleries along the inner walls connect to the Grand Jury Room between the two above the entrance hall, which is reached by a simple iron-balustered staircase.

However, looking up in the courtrooms unveils the remarkable truth: massive oak tie beams, collars, wind braces and purlins, and round-headed Norman windows in the Crown Court, reveal this to be, in fact, a huge medieval Great Hall. Moreover, it is thought to be the oldest surviving aisle and bay hall in Europe, the earliest residential timber roof in Europe (although much renewed later) and one of the largest medieval hall roofs of any type anywhere to survive. It was built by Robert le Bossu, the Second Earl of Leicester, in the mid-twelfth century. John of Gaunt added kitchens which have since disappeared, though the vaulted cellars remain, in urgent need of conservation. The assizes sat, as at the Great Hall at Winchester, at either end of the open hall until the 1821 remodelling. Further alterations were made by William Parsons in 1856, and this is what can be seen from the river on the far side of the building.

Leicester is no architectural masterpiece. The successive alterations and extensions are incoherent, often clumsy and rarely in themselves architecturally that successful. But these very characteristics help Leicester to speak more clearly about both the continuity and evolution of law courts than any other site. The startling and crude juxtaposition of Georgian court and Norman roof is an immediate and dramatic reminder that courts have sat in the same building for most of the last thousand years.

After the Crown Court moved to new premises in 1980 the building housed the County Court, but since that moved out too in the 1990s, it has been empty and unused. Nevertheless it is in excellent condition. The only exception is the fourteenth century cellars, which Leicester City Council hopes to repair soon. Sadly, the conservation, public opening and interpretation of the Great Hall have fallen victim to swinging cuts in the budget of the Museums Section of the council, which has been forced to close existing museums let alone consider opening new ones. But this remarkable building deserves an imaginative Lottery-funded scheme to open up the principal spaces to visitors and reuse some of the ancillary rooms for community or revenue earning purposes.

Lincoln County Hall
Castle Yard   Grade I

Sir Robert Smirke, best known for his magnificently Neoclassical British Museum, was equally prepared to design in a Gothic idiom when circumstances suggested it, as they did when he was invited to design a replacement for the old and decaying County Hall in Lincoln's Castle Yard. His two storey building was erected between 1822 and 1828, with Crown and Nisi Prius Courts for the assizes. It is little altered today. The façade spreads horizontally. The middle bays push forward through the ground floor arcade and there are prominent tall turrets in the corners. The interiors have splendid Gothic decoration, including ribbed plaster ceilings. The oak furnishings in the courts are superb, with panelled galleries and elaborate ogee canopies above the judges' seats. Public seating is in straight rows, but beyond the gangway the compartments of the well of the court form a tight semi-circle around the clerk's seat. The Grand Jury Room is now Court 3.

Although a Combined Court Centre opened in the city in 1991, the Lord Chancellor's Department has retained the County Hall for many criminal cases, thereby continuing a tradition of dispensing justice in the castle which probably dates back uninterrupted to 1068. With the PSA it undertook a comprehensive repair and refurbishment of the building in the early 1990s, which included the excellent restoration of the decorative plasterwork. The architects were Rodney Melville and Partners. Lincoln demonstrates that provision of modern facilities and adoption of the combined court centre concept does not have to mean the abandonment of historic court buildings.

Also worth note in the city are the restrained white brick Judge' Lodgings outside the east gate to the castle, built 1809-12, by William Hayward, who was also responsible in 1805-9 for the City Gaol and Sessions House on Monks Road. Some of the former and all of the latter, complete with all the original grained pine fittings, survives as part of the campus of North Lincolnshire College. The site had become a police station and was acquired by the college in 1993. The Georgian courtroom is used for occasional lectures, seminars and arts performances and it has proved particularly valuable as an aid on the tourism studies course to discussions about historic building management and interpretation.

Liverpool County Sessions House
William Brown Street   Grade II*

Facing Liverpool's majestic St George's Hall across William Brown Street is the city's second lavish and redundant law court. The Hall's architect, Harvey Longsdale Elmes, had planned a spectacular Daily Courts building for the site, but his scheme went unbuilt and it was not until 1882-4 that a courthouse was erected on the site, and then to the very different designs of George Holme, the County Surveyor, assisted by Francis Holme. Its ornate and foursquare free-classical exterior is matched by magnificent interiors on the piano noble: a delightful Italian Renaissance staircase hall with little saucer domes, tesserae floors and sgraffito under the ceiling; the rich oak, walnut and mirrored Magistrates' Room; the barristers' library; marvellous loos, with mosaic floors and original wooden-seated Shanks patent closets; and the two top-lit courtrooms, both with complete furnishings and richly plastered coved ceilings, the principle court very grand with superb panelling, the smaller court still with its ceiling gasoliers. Below is the witness waiting area and below that the glazed-brick cells; above is the grand jury room, kitchen and caretaker's flat. The planning is sophisticated, with segregated circulation linked to four different entrances, one on each side: magistrates and barristers up the stairs and under the portico, solicitors and witnesses via a pedimented door to the ground floor on the West side elevation, the public by a entrance in the utterly blank brick rear wall and defendants in vans and wagons direct into the basement, reached by a cartway enclosed by a high wall and iron gates on the East side.

The court closed in 1984 and became a museum of labour history. Few alterations were made and although the small Summary Court on the ground floor was cleared, its furnishings remain in store in the basement. After the museum was relocated, the building was taken over by the adjacent Walker Art Gallery, which currently uses much of it as storage and offices. However, it has long harboured ambitions to display more of its collection, currently in storage, in the building. Though as yet no plans have been formulated, and there is the small matter of fundraising, the principal courtroom will be the inevitable focus of conflict. It takes up so much internal volume that, whatever form the final proposals take, the Walker will almost certainly want to remove most of its furnishings, and probably install a false level floor across the well of the court in order to turn it gallery space with full disabled access. A less intrusive alternative would be to adapt the room as a lecture theatre. If it is converted in any form then the second courtroom should certainly be preserved intact.

However, such a course of action should be very carefully considered. The quality of the interiors of this building, and the completeness of the principal rooms, are such that there is a very strong case for restoring the piano noble, and the cells, to their original state as an exhibit in their own right. The Walker should focus on the possibilities of housing the majority of the new accommodation it desires in an extension built on the car park at the rear of the court. Only if it can be demonstrated that such an extension cannot provide an adequate amount of suitable floorspace should conversion of the best rooms in the Sessions House be considered at all.

101

St George's Hall, Liverpool

Lime Street   Grade I

One of the very greatest Neoclassical buildings in the world and a monumental statement of the mercantile prosperity and civic ambition of Liverpool. It was conceived as two separate buildings, a concert hall and the Assizes Courts, but after both competitions (in 1839 and 1840) were won by the same unknown architect in his twenties, Harvey Lonsdale Elmes, it was decided to combine the functions in a single building. This, however, was to be designed by the Corporation Surveyor, Franklin. Elmes protested and was finally awarded the commission.

His designs were ready in 1841, but he was killed by TB during construction at the age of only thirty three, and C R Cockerell and the engineer Sir Robert Rawlinson, both friends, stepped in to oversee the project through to completion in 1854.

Beneath the magnificent exterior the plan is simple: two courts flanking a massive vaulted central hall along the longitudinal axis, the idea being to create a vista 300 ft long from the bench of one courtroom to the bench of the other at the far end. Beyond the Nisi Prius court is Cockerell's exquisite Concert Room. This concept, however, could only be only achieved at the expense of functionality. For example, all rooms were reached from a single corridor running around the building and the only way to separate court users was to lock doors at strategic locations. Elmes' great vista was destroyed when Cockerell placed the huge organ across the north end of the hall, blocking off the Nisi Prius Court, and then when the screen to Crown Court at the other end was glazed after complaints from the judiciary about the noise. However, the interiors of both courtrooms, in keeping with the rest of the building, are magnificent, lined with granite columns and richly furnished in mahogany. The Crown Court has a barrel vault, the Nisi Prius a coved ceiling. The fittings remain intact in both.

The courts remained in use until 1984 (when one of the first combined court centres opened in Derby Square in the city), although the accommodation in the Hall, designed for occasional Assize sittings, was by then totally inadequate for the continuous sitting of a modern Crown Court. These were in fact pioneered here from 1956, such was the weight of business in Liverpool. The Lord Chancellor's Department had paid 97% of the running costs when the courts were in use.

When they closed, Liverpool City Council, accustomed to paying only 3%, refused to pick up the tab and for over a decade the building stood largely empty and decaying. But then the National Lottery arrived, offering a way to fund the investment required to repair and refurbish the building, and the council formed a trust to raise money to restore the building for all kinds of events and performances. Currently costed at £15m and designed by Purcell Miller Tritton, the key to the viability of this scheme is the refurbishment of the Cockerell concert room for performances and conferences and, with it, the creation of adequate catering facilities. It was planned to put these in the Nisi Prius Court by stripping it of its fittings, something unacceptable in a building of such importance, and certainly not something that should be funded by the Heritage Lottery Fund. Happily the trust responded to these concerns and the catering facilities will now be located elsewhere in the building so that both courtrooms will remain unaltered and will be restored in all their glory. Work has now begun and the city hopes to have completed the entire scheme by 2008, when it will be European Capital of Culture.

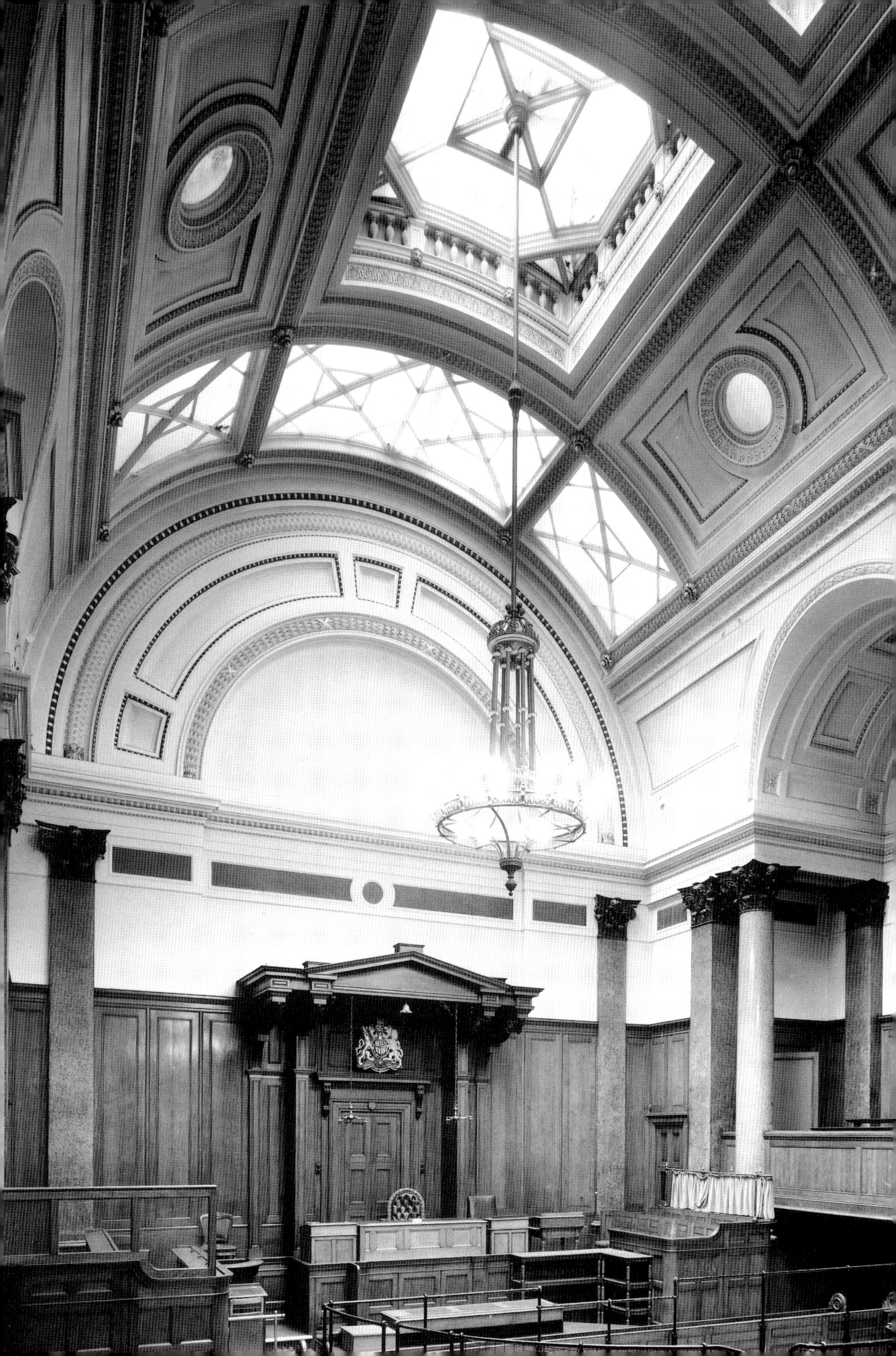

London:

Great Marlborough Street Magistrates' Court
Great Marlborough Street, Westminster — Unlisted

West London Magistrates' Court
Southcombe Street, Hammersmith — Unlisted

Tower Bridge Magistrates' Court
Tooley Street, Southwark — Unlisted

Old Street Magistrates' Court
Old Street, Hackney — Unlisted

Clerkenwell Magistrates' Court
King's Cross Road, Islington — Grade II

These five London Magistrates' Courts were built as Police Courts by John Dixon Butler, architect to the Metropolitan Police, between 1905 and 1914 (Clerkenwell also has an earlier phase, dated 1841-2). Old Street and Great Marlborough Street were also police stations. In total Butler built nine courts as part of an expansion programme launched by the Met after it took over responsibility for Police Courts in London in 1897. They are immediately identifiable, distinguished by a sophisticated free classicism executed in red brick and Portland stone. One particularly distinctive motif is the elegant, attenuated brackets supporting pediments over windows and doors. Some of these buildings, such as Old Street and Tower Bridge are really quite grand, with tall, shallow frontages and impressive pedimented entrances. The interiors are equally distinctive. Entrance halls have mosaic floor tiles with the MP monogram and

Old Street Magistrates' Court

Old Street Magistrates' Court

104

d Street
agistrates' Court

often they are top-lit with barrel vaulted ceilings. Many have good staircases with handsome iron balustrades (as at Old Street). The courtrooms have cleanly designed oak furnishings. Witness boxes have testers, docks elegant iron railings and the raised bench is often backed by a curved screen. They are top-lit too.

Now these handsome buildings are falling victim to the drastic rationalization of court facilities in London – Great Marlborough Street, Old Street, West London and Clerkenwell have all closed, Tower Bridge will soon – and attention is turning to their adaptation and reuse. Permission has been granted to convert Great Marlborough Street into a hotel. The courtroom will become the restaurant; Westminster City Council want as many of the fittings to be retained as possible. West London is the subject of an excellent scheme by Pawson Williams Architects to convert it into offices. An attractive courtyard will be created between restrained new wings erected in the existing yard behind the building. The courtroom will become a seminar/reception room with the dais and witness box retained. Because the new courtyard wings will provide open plan office space, the client has no need to alter or open up the cellular offices and apartments in the Dixon Butler building, which will be sympathetically refurbished and reused. Most recently Levitt Bernstein completed a feasibility study into the potential of turning Old Street into a community law centre featuring a business law clinic and the reuse of the courtrooms as seminar and training areas.

By and large these schemes seem pleasantly imaginative, but it is important that at least one representative courtroom interior is preserved in one of the buildings intact.

West London Magistrates' Court

Great Marlborough Street Magistrates' Court

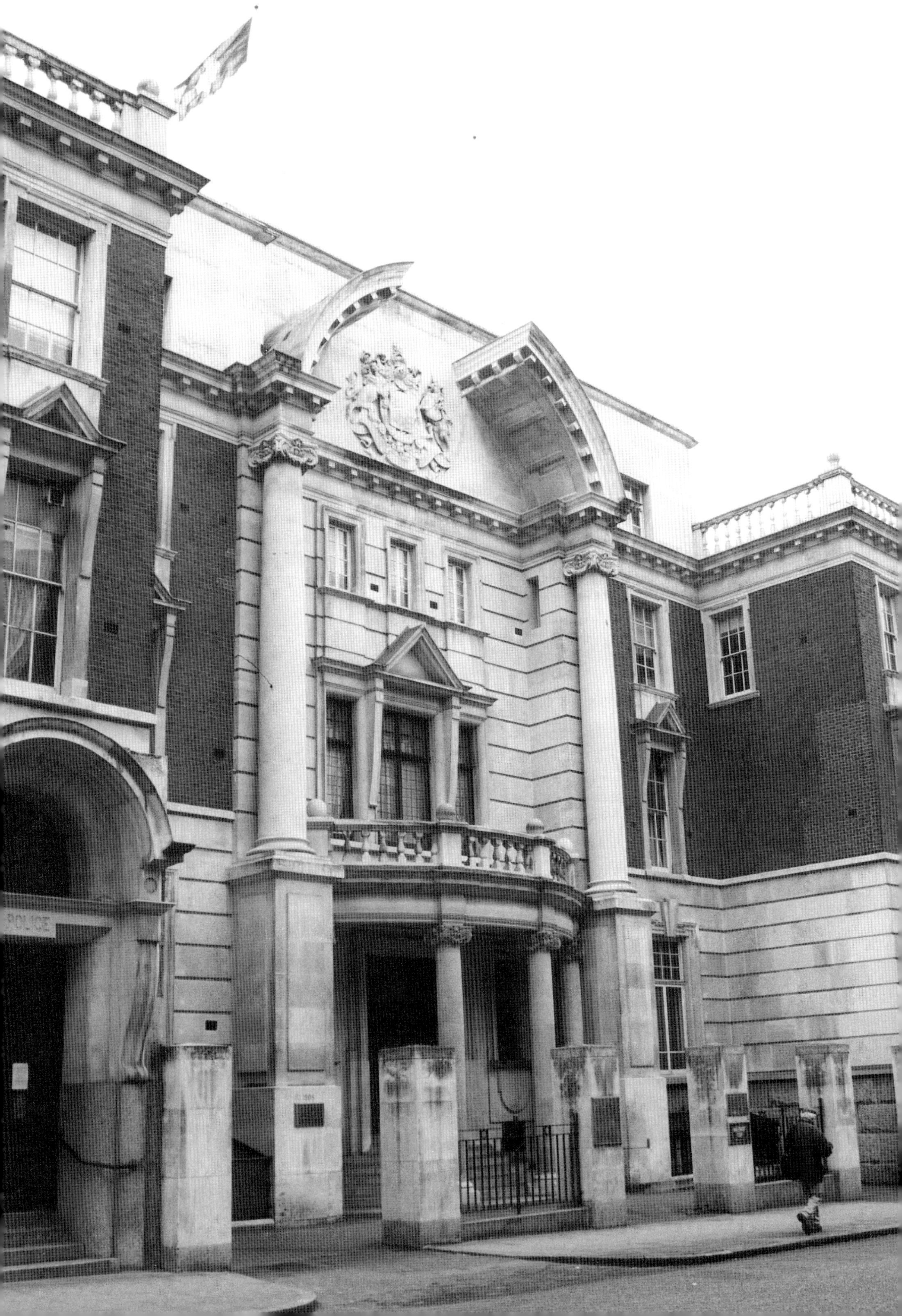
POLICE

Lambeth Police Court, London

Renfrew Rd, Southwark Grade II

This Tudor Gothic Police Court by T C Sorby, architect and surveyor to the Metropolitan Police, was completed in 1869. It formed part of a mini civic centre with a fire station (now apartments), a police station and a workhouse, part of which remains. In the centre of Sorby's building was the double-height courtroom, with offices to one side and the custody range, with cells on two storeys, to the other. Towards the end of its career as a Magistrates' Court (it closed in 1990) it was used for arraignment hearings of terrorist suspects and fitted with high security features such as bullet proof glass and a lead-lined floor.

After closure it was bought by a developer who intended to create nine apartments within it, but a vocal local campaign, arguing for community use, succeeded in persuading councillors to refuse permission. Nevertheless, at appeal, and subject to modifications to the scheme, the developer got his permission and put the building up for auction in 1995, where it was bought by the Jamyang Buddhist Centre. They believed the courthouse would make a wonderful location for their programme of meditation, healing and learning. They had previously made an offer for the building when it was first put on the market and had joined the campaign against the apartment scheme. Since acquisition the centre has undertaken phased repairs, with the help of Lottery funding, to convert the building to its new use. The roof in particular was in a terrible state.

And Jamyang was right – a Police Court appears curiously well suited for life as a Buddhist community centre. The courtroom, spacious and top-lit, remains the natural heart of the building, only now as a Shrine Room in which a seated Buddha occupies the dais under the canopy where the magistrates used to sit! The holding cells have become meditation cells and accommodation for friends and volunteers. A café has been opened, spilling out onto a peace garden created by replanting the old yard. Throughout, a light touch has been shown – doors, fittings and signwriting are preserved in place, alterations kept to a minimum. Much of the work has been undertaken by volunteers. Jamyang's achievements were celebrated in May 1999 when the building was blessed by the Dalai Lama.

The Jamyang Buddhist Centre is as a refreshing example of how imagination, care and commitment can breath a positive and public new life into a redundant civic building.

Willesden Magistrates' Court, London St Mary Road, Harlesden

Grade II

Another empty former Magistrates' Court, this one designed by H T Wakeham for Middlesex County Council in 1899 and extended by G Goodson & Sons in 1923. Free Tudor. Red brick with stone dressings. It closed in 1989 and is now in a deplorable condition. Harlesden has well known problems, but it ought to be possible to find a use for this building. There is nothing particularly significant about the interiors.

Lutterworth Magistrates' Court
Gilmorton Road   Grade II

Another charming Magistrates' Court that has fallen victim to the cost-saving contraction of rural Magistrates' Courts. Lutterworth was built in 1910 next to the police house and superintendent's house of 1847, on a triangle of land between two roads on the edge of the town. It is a building of modest, but impressive quality, an exercise in a slightly Lutyensesque classical idiom. The high hipped roof is topped by a louvered cupola, and balanced by flanking doors and doomed vestibules. The quality is evident in the detailing – such as the bracketed eves cornice – and the workmanship. This carries over into the unaltered courtroom interior, dominated by the substantial and excellently finished oak fittings and panelling. Elsewhere many other original fittings survive, such as the art nouveau tiles and marble basin in the judge's loo.

The court closed in 1998 and was put on the market. Extremely solidly built, there can be few structural concerns. However, alternative uses raised the same vexed questions faced at many other courts. Aside from the courtroom, the building contains few other large spaces, mostly corridors and small cellular rooms. The courtroom fittings are complete, but many productive new uses would require the removal of most of them. In fact the eventual purchasers proposed a sensible compromise in order to adapt the building for office use – clearing the public benches to make way for desks etc., but leaving the dock, the well of the court, the witness box and the bench unaltered. This kind of low intensity use is an intelligent response to the apparently conflicting demands of conservation and viability in many courthouses, a demonstration that the two need not be in opposition.

Maldon County Court
London Road   Grade II

Built 1858 in grey Gault bricks with ashlar dressing, this must be by Charles Reeves. It was erected right in the middle of his tenure of the post of Surveyor of County Courts and is absolutely typical of the single-storey version of his standard Italianate design. The bracketed cornice and free-standing stone Royal Arms above the parapet are characteristic. The symmetrical seven-bay principal façade has entrances at either end, one for the judge and one for everyone else. A public office and offices for the court officials occupied this front section, behind which is the double-height courtroom.

The building was doubling as the Magistrates' Court when it was closed in the early 1990s. It was put on the market and bought by a local couple in 1995, despite the best efforts of the Lord Chancellor's Department which received three successive offers from them, each lower than the last, before it finally accepted the last, and lowest! Since then the new owners have heroically set about turning it into a family home. Heroically, because they are doing it mostly on their own and because by the time they bought it was in a pretty terrible condition. Already worn out after years of under-maintenance, whilst empty its condition had deteriorated to such an extent that the courtroom was beginning to resemble a conservatory, such was the extent of plant growth. It had fifteen loos and a urinal, but no hot water, and ancient electrics. The conversion has been a long-haul, but all that effort has created a spacious family home with a spectacular central living space in the old courtroom. Here, where only the dais and the witness box survived, the new owners have maximized the use of the space by creating a mezzanine around three sides. Glazed doors open out onto a lovely little courtyard in heart of the building, and a spiral staircase gives access to a new terrace on the roof above with wonderful views over the town to the estuary of the River Blackwater to the east. The hard work has been well worth it.

Minshull Street Crown Court, Manchester
Minshull Street
Grade II*

Minshull Street is the great standard bearer, the often quoted example of what it is possible to do with an historic court building. And rightly so.

The building was completed as the City Police Courts in 1871 to the designs of Thomas Worthington – a solid piece of Italian Gothic, with a massive tower. It was after Alfred Waterhouse's greatly admired Assize Courts were destroyed in the Blitz that the Assizes moved in. As completed it had four courtrooms; another was later created, but by the mid 1980s these five were totally inadequate and a study was undertaken in 1987 to see if it would be possible to expand capacity on the site to eleven courtrooms plus all the associated accommodation by then expected of a major Crown Court centre.

The Hurdrolland Partnership have triumphantly proved that it was. The Victorian courthouse was U-shaped around an open courtyard. The key to the Partnership's scheme was an empty warehouse which closed this courtyard. This was replaced with a new wing, including a new main entrance, and the courtyard was glazed over to create an impressive atrium that serves as the central circulation space uniting serving both old and new wings. The four original courts were refurbished, with suspended ceilings removed to reveal the original Victorian top-lighting. Six new courtrooms were created. The greatest design challenge was to meet the complex circulation requirements in a readily understandable form. This was achieved by vertical separation: defendants on remand arrive at basement level; the public and legal professionals arrive at street level and circulate around the atrium on ground and first floors, with direct access to the court rooms; jury members use a mezzanine floor and the judiciary the level above that, each again with separate entrances to every courtroom. The separation is expressed by the balconies around the atrium. Whilst the major Victorian spaces were restored and the secondary ones reordered, the new building and the atrium were treated as unashamedly contemporary elements. Externally, however, the decision was taken to dress the new wing in a pared down interpretation of Worthington's design, treating it as an additional gabled bay to the original pair. Pastiche has been avoided by careful attention to the rhythm and proportions of the original building. The architect, Jim Stevenson, discusses his approach in the Introduction.

The building has proved to be very popular with its users, particular judges, who believe that the grandeur of the refurbished Victorian courtrooms, by lending gravitas and even a certain amount of awe, adds to the quality of justice dispensed. Minshull Street demonstrates powerfully that listed courthouses can provide efficient, effective and popular court facilities, with a bit of extra land and plenty of imagination.

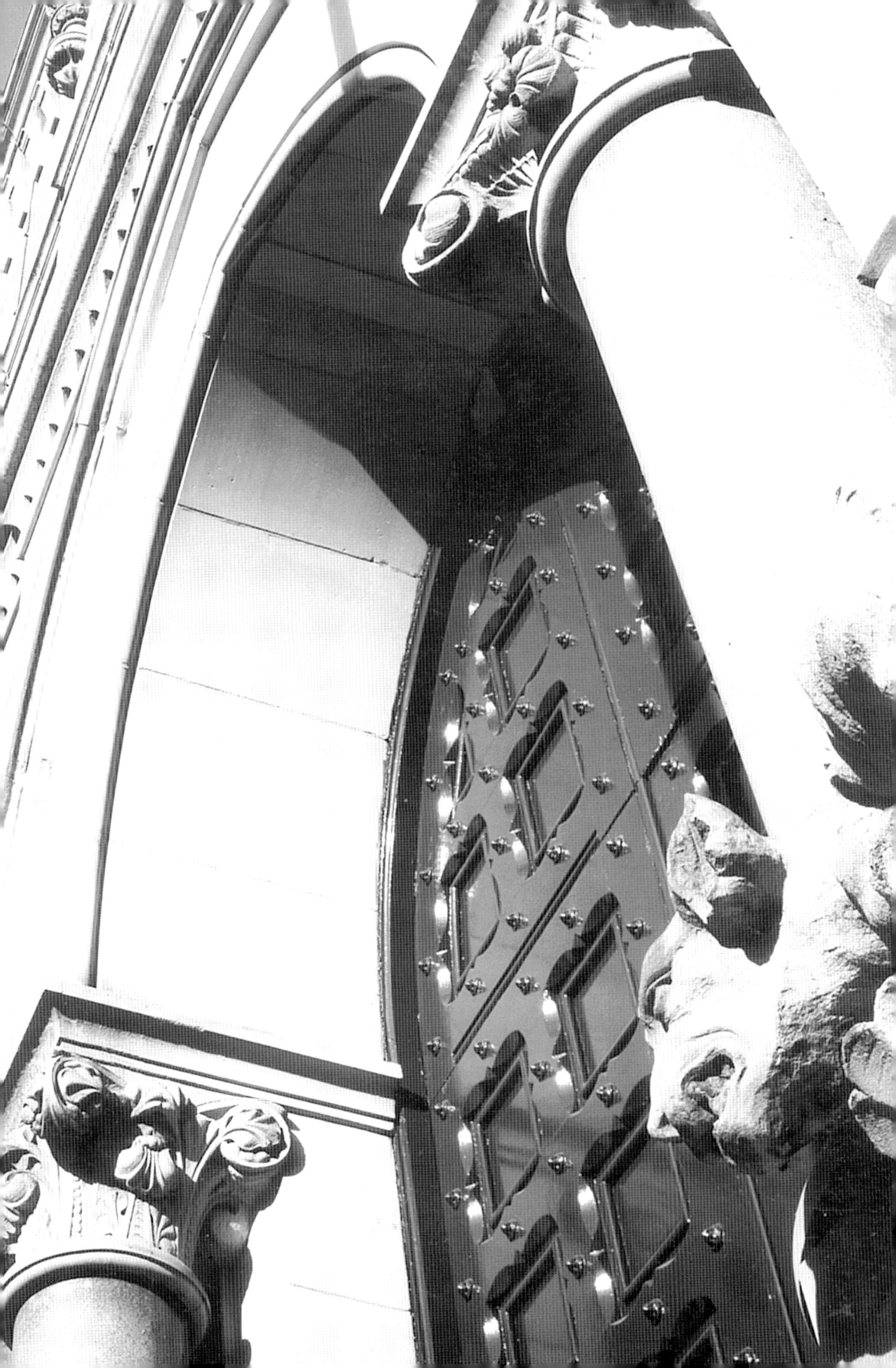

Manchester Magistrates Court
Crown Square   unlisted

Very few postwar law courts stand up to much architectural criticism. On the whole, the great building programme that followed the abolition of the assize system in 1971 – despite constituting the most significant group of public buildings of the last twenty five years – only serves to illustrate the depths to which most public architecture in England descended during the 1970s, 80s and 90s. But there are exceptions, and two of these are earlier Magistrates' Courts. One, Chesterfield, is listed (q.v.); the other, Manchester, is not, but deserves consideration.

It was completed in 1971 and designed by Yorke, Rosenberg & Mardall. It is a disciplined, formal building, classical in conception – a vast pavilion raised on piloti above a podium. It forms one side of a now slightly shabby public square, along with The Guardian newspaper offices and the 1950s Crown and County Courts (not nearly so successful). The frame, a powerful but controlled grid in front of recessed glazed walls, is clad in white tiles, the characteristic motif of the practice at the time. The frame clearly expresses the internal planning: the offices one side and the accommodation for the magistrates and lawyers on the other are marked by a module one storey high; the courtrooms through the middle by double-height modules. The result is like the tripartite division of a portico'd Palladian façade. The interior reveals YRM's control over detailing – travertine was used to clad the double height public spaces and Charles Eames seating was chosen for the waiting areas. Sadly, though, the clarity of the design and the plan are not as evident today because the entire building is covered in netting to protect pedestrians from falling tiles, vivid evidence of the lack of maintenance from which the building has suffered.

Now, under a PFI scheme, the building is to be demolished and replaced by a new Magistrates' Court designed by Gensler, which is being built on an adjacent site previously occupied by the offices of The Guardian newspaper. The site of the YRM building will be redeveloped with offices as part of the financing of the scheme. These developments are seen as a lynchpin of a massive redevelopment of the area, which is being marketed under the name Spinningfields. Denton Corker Marshall's new Civic Justice Centre (see Chapter 2) also forms part of the masterplan.

Demolition is welcomed by many. One employee described it as an 'ugly lavatorial mistake'. But SAVE believes that these sentiments are misguided: the Gensler scheme appears very unlikely to offer any improvement for the public realm over the refurbishment of the YRM building, which would undoubtedly have revealed the underlying elegance of the structure. HM Court Service Inspectorate has stated that 'the court buildings present serious problems of design and maintenance' and that the 'courthouse itself presents many restrictions'. SAVE has heard such comments many times before about many different buildings, and very often they have been shown to be unfounded or at the very least overstated. No doubt similar comments were made about Derby Shire Hall, until the recent will and way was found to refurbish it under a PFI scheme as another large Magistrates' Court centre (q.v.). But sadly the failure of the Twentieth Century Society's attempt to get YRM's building listed appears to have sealed its fate.

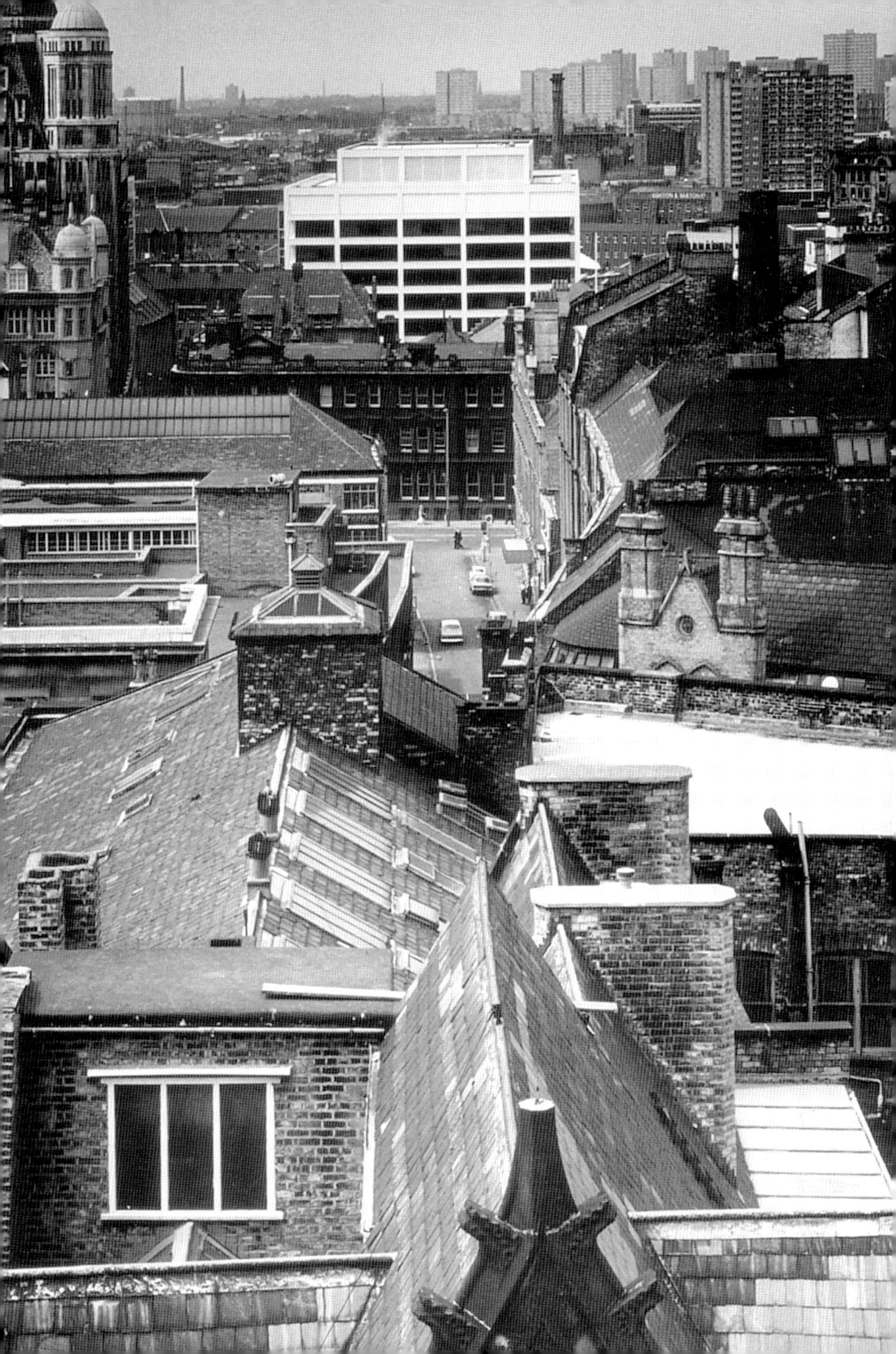

Morpeth Sessions House
Castle Bank   Grade II*

If there was a Sessions House style in the early nineteenth century it was that typified by Beverley, Wakefield and Devizes: classical, pedimented, possibly porticoed, with lower wings flanking the central courtroom. But there were always alternatives, and Morpeth (1822-8) is one of the more original, a massive medieval gateway to the gaol behind (semi-radial, demolished 1891), some 72 ft high, with corbelled battlements, a vaulted passageway through the centre and an eight-sided apsidal back. The designer was John Dobson, the leading North East architect, responsible for Newcastle Central Station and much else. Most of this was classical, but Dobson was a capable Gothic architect too. He said that Morpeth was inspired by Caernarvon, Conway and Beaumaris Castles, but he was also working within an established castle-style tradition in the northern counties which had become fashionable in the early nineteenth century following a surge of interest in the great feudal barons of the medieval north and their castles.

In 1981 the magistrates moved out. They were far from happy with the limited facilities in the building and the owner, the Northumberland Police Authority, was concerned about the structural condition of the building, which it was estimated would cost £320,000 to put right. It applied for consent to demolish the building, but the proposal was fought by the District Council and local campaigners determined to save this cherished local landmark, which stands guard on the old Great North Road as it enters the town from the South. Permission was refused, grant aid promised and a scheme put forward to turn the building into flats, arranged around a semi-circular courtyard created in the heart of the building by taking out the courtroom itself. In the end, a small number of flats were created, but the courtroom was retained, as a restaurant.

We should be thankful for the determination of the local community because Morpeth is a wonderful building, with one of the grandest interiors of any Sessions House. In the left-hand tower there is a magisterial imperial staircase, all in stone, rising to a first floor vestibule, under a vaulted ceiling and lit by a tall mullioned and transomed window. An arcaded first floor gallery runs along one side. Off the vestibule, in the right-hand tower, is the handsome magistrates' room with moulded ceiling; at the rear is the apsidal courtroom. This is similar to and no doubt inspired by Thomas Harrison's 1788 Shire Hall at Lancaster Castle (q.v.), a very grand space with a public gallery running around the seven sides of the apse on top of a stone arcade and an elaborate rib-vaulted plaster ceiling. Most of the semi-circular court fittings, including the dock, remained until closure; now only the bench's canopy survives (with a minstrel's gallery above).

The courthouse has recently changed hands and is well maintained, even if the interiors are painted in rather lurid colours. The new owners have converted it into a women-only health centre. Fitness machines now line the public gallery and the well of the court is an aerobics floor, with mirrors in the arched recesses underneath the gallery. As an alternative use this is surprisingly successful. Above all it retains the integrity of the principal spaces; there are no subdivisions. The most insensitive alteration is the boxing-in of the bench, now fronted by full-length mirrors. Only the crotched peak of the canopy is visible. Behind, tanning booths, bathed in a weird neon light, have been installed in place of the magistrates' seats. What Victorian JPs would have made of this goodness only knows.

Northampton Sessions House
George Row   Grade I

On 20th September 1675 the Great Fire of Northampton swept through the midlands town. Among the many buildings destroyed was a Sessions House, either the temporary 'board and timber shed' erected only five years before, or the beginnings of a new permanent structure for the assizes intended to replace the Castle, in which they had sat since the Middle Ages, which was demolished in 1662. Following the fire it was decided to erect a new Sessions House on part of the site of the destroyed gaol, on what is now George Row. The building that arose was one of the most architecturally sophisticated courthouses of the century, and it is now the only Assize Court of the period to survive with original fittings and panelling. Today it lies at the heart of a seventeenth, eighteenth and nineteenth century civic ensemble, also including County Hall, the Judges Lodgings, holding cells, gaol buildings and a hanging yard.

The most likely architect is Henry Bell of King's Lynn, who was charged with what would now be termed masterplanning the rebuilding of the town and to whom All Saints Church is also attributed. The most precise physical evidence are the idiosyncratic window frames, with acanthus scrolls flanking keystones with masks, which can be found on other Bell buildings. The principal north facade is wonderfully vigorous Restoration stuff, one of the few surviving public buildings of the period. The plan is unusual. Most seventeenth and eighteenth century Assize Courts, such as Derby, York and Bedford, were in essence a single hall with courts at either end, open or only partly partitioned, which lent themselves easily to wrapping in a symmetrical facade. At all these buildings the courts were later either moved or fully partitioned from the hall because of complaints from the judiciary about the difficulty of conducting two trials simultaneously in the same space.

Northampton appears to be an early attempt to address these problems. The two courts were arranged one in each arm of an open L shaped hall, with the judge's benches at the ends of the wings. In this way proceedings in one court were less likely to be disturbed by happenings in the other, hidden around the corner. Marrying this arrangement to a symmetrical facade on George Row created unusual ceremonial routes (the arrival of the Assize judges, escorted by the Lord Lieutenant and his javelin carriers, was one of the great public occasions of the county calendar). The left-hand door led directly onto the dais of the Nisi Prius Court and the right-hand one opened into the vestibule that separated the two courts, in line with the judge's seat at the far end of the Crown Court immediately ahead. Both these great pedimented doors are crowned by royal arms, unlike the public entrance on the west facade. The Crown Court was not partitioned off until 1812 and the Nisi Prius until even later (and far more crudely).

Both courtrooms are complete with the full array of panelling, fixtures and fittings. Some of this is original, though over the centuries there has been extensive reordering and replacement in sympathetic Neo-seventeenth century style. Nevertheless both interiors, particularly the Nisi Prius court, remain rich in seventeenth and eighteenth century spirit and form. Both have splendid seventeenth century candelabras and the Nisi Prius also retains a wonderful set of full length portraits of monarchs. But the chief glories of the interior are the magnificent plaster ceilings by Edward Goudge (described in 1688 as the 'beste maste in England in his profession') installed ten years after the rest of the building had been completed (possibly because such finishing touches were delayed so that the building could be brought into use as quickly as possible to deal with the great volume of legal business created by the fire). The plasterwork is sumptuous and full, rich in the acanthus scrolls and garlands of fruit and flowers typical of the date, but also very specifically designed to reflect the function of the building. Thus over the body of each court is an enriched arms of England, Scotland, Ireland and Wales, and the schemes get steadily richer as they approach the judges' seats. Above each dais is a putto holding the scales of justice; in the Crown Court he is flanked by an assortment of chains, leg irons and handcuffs and in place of one of the Nisi Prius's angels is a devil, whose tongue, it was said, wagged at the sound of false evidence.

By the 1980s facilities were unsurprisingly considered pretty basic: in addition to the building described there was only two jury rooms, two judge's retiring rooms, a robing room, the cells below and the 'blue room' (created out of a small yard). Two other courtrooms were provided in austere fashion in the old Co-op Furnishing Rooms elsewhere in town. So a new Combined Court Centre was built and when this opened in 1986 the Sessions House finally closed. Since then it has been maintained in good condition by Northampton County Council, which opens it for Heritage Open Days and occasional meetings. But so far a long-term strategy for the building has yet to emerge. The County fully recognizes its architectural importance and has expressed a commitment to finding sympathetic uses which allow for public access. The most obvious solution would seem to be as a venue for meetings, talks, seminars and lectures, both for the County and other groups, but progress will require an injection of urgency and finance.

Nottingham Shire Hall

High Pavement   Grade II*

Some local authorities and town councils have found that once declared redundant, certain historic law courts can be conserved and opened to the public at relatively low cost as minor tourist attractions, explaining the judicial process of two centuries ago, or perhaps describing a notorious local trial. But these schemes are limited in their scope, essentially a quick and easy means of making some use of an architecturally important – and listed - building or suite of rooms that the owners are constrained from converting to more profitable use. The Galleries of Justice at Nottingham's Shire Hall, however, are quite different in their scope and ambition. They were conceived as a national museum of law, the brainchild of a Reading solicitor, Geoffrey Goldsmith, in the 1980s. His search for a location that would bring together all the strands of the subject - penal detention as well as the dispensation of justice - brought him to Nottingham, where the Shire Hall was lying rotting after the Crown Court had left in 1987 and a subsequent hotel conversion scheme had failed. In 1990 the reuse of the empty building was identified as a key objective in the emerging regeneration strategy for the Lace Market area and the City Council, to its great credit, stepped into buy it on behalf of the trust established to realize the Galleries of Justice idea. The museum opened in 1995, since establishing itself as a popular visitor attraction and one of the lynch pins of the revitalized Lace Market. The story of justice is told through displays and interpretation both in the Shire Hall and the gaol complex behind, which includes medieval, eighteenth and nineteenth century cells and prison buildings, including the Governor's House and an Edwardian police station.

The ultimate aim is official recognition as the National Museum of Justice.

As to the architecture, the sophisticated Doric façade on High Pavement is by James Gandon (1770-2), architect of Dublin's Custom House and Four Courts, but little else of his Shire Hall survived a fire of 1876. The projecting stair turret to the right and the interior, including the arcaded central hall and the Crown and Nisi Prius courts, are by T C Hine immediately afterwards. The impressive courtrooms have public galleries on three sides behind colonnades, and richly plastered ceilings. The oak fittings are complete and unaltered. The gaol behind includes remains from many successive phases, the earliest dating from about 1618.

Oakham Castle

Grade 1, Scheduled Ancient Monument

Down an alley off the market square in Oakham and through a thirteenth century gateway topped with a pediment added by George Villiers in 1621, lies Oakham Castle, one of the most romantic and historic of all working courts. Today, almost all that is visible is the stone Great Hall, standing entirely alone in a grass enclosure bordered by trees on one side and the back of the town on the other. This remarkable building, built by Walkelin de Ferrers in c.1180-90, is the earliest hall in any English castle to survive so completely. The exterior is, of course, much restored but inside are arcades of four bays with circular piers and beautifully carved capitals. The castle is chiefly known now for the remarkable collection horseshoes that cover the walls of the hall; since at least 1521 the Lord of the Manor has required every peer to forfeit a horseshoe on his first visit. However, it has also been the site of a court since at least 1229 when the first Assize was recorded. The last Assize sat in 1970, but a Magistrates' Court is still held here.

Like the choir in a church, a charming little group of late Georgian court furnishings are arranged on a raised platform at the west end of the hall facing the bench set against the west wall with a pediment above judge's chair. The court is entirely open to the rest of the hall, a survivor of the arrangement normal until the end of the eighteenth century, when the increased professionalization of the law led to the full enclosure of most courtrooms to avoid disturbance. Similar fittings at the east end were removed in 1911. The magistrates now sit in an extension built in the early nineteenth century as the jury room. Despite a suspended ceiling this retains some good original joinery and features some good post-war wooden court furnishings.

Sadly, though, the judicial tradition stretching back nearly eight hundred years will probably not survive much longer. Despite a fierce rearguard action being fought by the Oakham bench and the local community, the closure of the court seems likely. Facilities have been declared inadequate and when the local bench failed in its bid for refurbishment funding, it was decided to merge it with Melton Mowbray. Many fear that this is a prelude to the closure of the court. It only sits once a week, something frowned upon by the Lord Chancellor's Derpartment as inefficient. Communities today must be far more isolated than Oakham before the public service benefit of a local Magistrates' Court is deemed by central Government to outweigh its cash-savings-driven enthusiasm for closing small, lightly-used courthouses.

Oldham County Court
Church Lane   Grade II

Like Warrington (q.v.), this is a County Court by Sir Henry Tanner and the family resemblance is clear: French gothic in red brick and buff terracotta, with some crisp attractive detailing, the first floor court evident from the tall hipped roof and prominent gabled windows. Sadly the attractive cupolas are lost. It occupies a pretty location in a narrow street of largely eighteenth century houses immediately below St Mary's parish church. The date is 1894 and the building closed in the 1990s. Happily it has found another bustling community use, this time as the Salt Cellar Community Church. Once again the flexibility and space in these buildings has been well utilized. Downstairs is a café, upstairs the lofty courtroom is used for worship, concerts, youth groups and many other activities, though only the dais and wainscoting remain.

Oxford County Hall

Oxford Castle, New Road				Grade II*

The assizes sat in Oxford Castle until the Black Assize of 1577, when Prison Fever claimed the lives of the judge and the jury. They were promptly transferred to the Guildhall where they remained until – in a story familiar to other towns and cities – complaints from the judiciary about conditions there led the magistrates to build new courts, in this case back at the castle in a new County Hall erected in 1840-41 to the designs of John Plowman. The castle had always been the town's prison and had been rebuilt in the late eighteenth century by the leading prison architect of the day, John Blackburn. Therefore, as at Canterbury and Durham, prisoners could now be brought to the dock direct from the prison via an underground passage.

Plowman's design was in keeping with its surroundings – Neo-Norman with tall round-headed windows, the central bay pushed up above the porch and crowned with bartizans. It was originally flanked by curved walls terminating in drum piers enclosing the forecourt; today only the right-hand survives. Inside the planning is simple: the porch leads into the high top-lit hall with pilasters and a flagged floor and the usual display of portraits of the county's great and good; at the rear is the Grand Jury Room (now a kitchen with inserted floor); to the left is the Nisi Prius Court, to the right the Crown Court. The latter retains virtually all its fittings – throne with canopy, well, dock, jury box etc., and a public gallery. The Nisi

Prius courtroom has been used by the County Council as a council chamber since 1912 and substantially altered, although the Assizes sat here as well until 1972. The Crown Court was transferred to a new building in St. Aldgate in 1995 and since then the old Crown Court has found limited use as the Coroner's Court.

The future of the County Hall is tied up with the redevelopment of the rest of the castle site. The prison, occupying an outstanding collection of gaol buildings – some listed Grade I – and the Norman St George's Tower, was closed in 1996. Since then the County Council with the developers the Trevor Osborne Property Group have begun a comprehensive £37 million development that will allow public access to the castle, including the Norman motte, for the first time in centuries. The scheme, due for completion in 2004, involves the conversion of the various prison blocks into a four star hotel and a heritage centre and the construction of new buildings housing educational facilities, apartments, shops, bars and restaurants and leisure facilities. Architects involved include Dixon Jones, Panter Hudpsith, Richard Griffiths Architects and Architects Design Partnership.

The County Hall is the only part of the site not to form a part of the current scheme; what role it may play in the future has yet to be finalized. There is no reason why the County Council should not carry on using the old Nisi Prius court as a council chamber – this would be a welcome element of continuity. The Crown Court, as well preserved as it is, could form an extension of the new castle heritage centre. Sadly one proposal to send visitors down from the dock and through the tunnel to the prison was scuppered by the Health and Safety Executive. The front runner at the moment, though, is for some kind of conference/reception centre use (which, unfortunately, would require the retention of the kitchens in the Grand Jury Room).

Incidentally, the City of Oxford retained a separate Assize until 1972. With the opening of the new court centre in the 1980s its well preserved but underused late Victorian courtrooms in H T Hare's 1897 Town Hall have earned the City Council a useful income as the set for TV dramas, notably Inspector Morse. This has been the case at a number of other courts that form part of town halls, such as St Albans and Kingston. The new court building is one of the more successful of the combined court centres – regular users put that down to the influence in the design process of an effective users advisory panel.

Presteigne Shire Hall

Presteigne Grade II*

This is arguably the most remarkable survivor of all UK court buildings and a visit today gives the most complete picture of a Georgian courthouse it is possible to find.

Presteigne was the county town of Radnorshire - the least populated of the old Welsh counties – until 1888. The Shire Hall was built in 1826-9 to replace a seventeenth century building which had become increasingly derelict. Its rather provincial Neoclassical facade must then have seemed sternly grand and imposing, and it remains the most distinguished building in Broad Street, standing forward of neighbouring buildings in an uncompromising assertion of Crown rule in this remote part of the kingdom.

The architect was Edward Haycock, of a Shrewsbury firm of architect builders responsible for many local country houses. His estimate was £4,800. For that the local justices got what one nineteenth century judge described as the most comfortable judge's lodgings in the country and a perfectly formed, though tiny, courtroom. The importance of satisfying the sensibilities of the judiciary should not be underestimated. The celebrated case of Guilford in 1860 demonstrated that if the facilities weren't up to scratch the judges would threaten to take the Assizes elsewhere, with all the loss of prestige and business that would bring to the town.

The building is laid out either side of the courtroom, which projects into the street behind pediment and pilasters. To the left are the offices of the Clerk of the Peace, the public entrance and the Grand Jury Room; to the right the judge's lodgings. The beautiful courtroom, with vaulted ceiling and delicate plasterwork, is a perfect example of Georgian court planning, the various actors in the drama seated in segregated benches rising around the elliptical baize covered table. The accused entered directly from the cells below. This is an

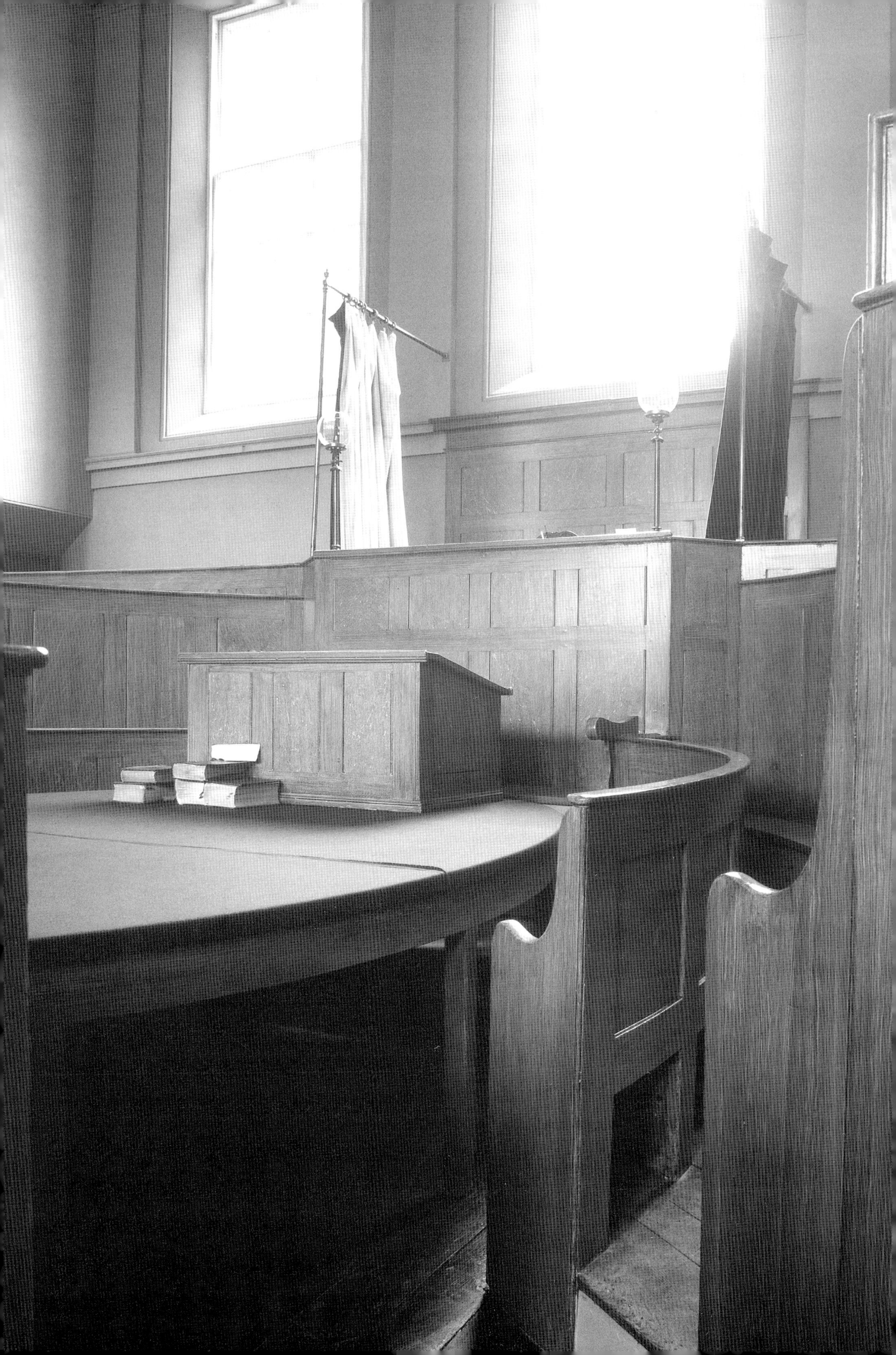

intimate space that must have been charged with energy when filled with officials, defendants, lawyers, jurors, witnesses, reporters and, of course, spectators. Here, perhaps more than anywhere else, you sense Georgian law as theatre; the tight and tiered benching, with the judge enthroned at its peak, imparts the majesty of the law despite its modest dimensions. Assizes were held here right up until 1971, after which the building became the local library and museum. Recently, thanks to EU and Cadw grants, the building has been magnificently restored. Not only has the courtroom remained virtually unaltered since construction – complete with a unique working gasolier of 1860 – but more remarkably the judge's lodgings have survived with 70% of their original furnishings. The exemplary quality of the restoration has revealed the original interior and exterior paint schemes, mid-nineteenth century dinner services, dressing and bedroom furniture, bathroom fittings and the full array of kitchen pots and pans. Much of this had been bought or commissioned especially for the building.

All this was made possible by the actions of the last caretaker, Mrs Rowe, who carefully stored the contents in the attic after the last Assizes rose in 1971. Dr. Charles Kightly, the project director, deserves plaudits too for persuading Powys County Council to restore the building after he learnt of this remarkable secret. The building is now open to the public between May and October.

Ripon Court House
Minster Road

Grade II*

The closure of this delightful Georgian Magistrates' Court was described by one magistrate as an 'act of stupidity and barbarity'. The present structure replaced the medieval Liberty Court, demolished in 1830. The Ripon Liberty (once called Riponshire) originates from a grant of territory to St Wilfred in 661. It was subsequently passed down to the Archdeacons of York, under whose courts the Liberty was governed until 1888 when his Court of Quarter Session was absorbed into the West Riding. The amalgamation of the Ripon Liberty Division with neighbouring benches to form a new Harrogate Division in 1998 brought this remarkable history to an end.

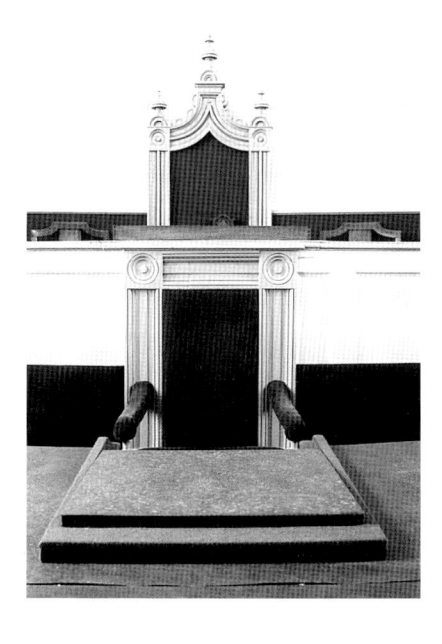

The building, Pevsner rightly says, looks like a nonconformist chapel, with a hipped roof and dressed stone and ashlar dressings. The main entrance, under a pediment supported by Doric columns, leads into a cross passage. To the left lie the Old Jury Room and the Justices' Retiring Room, to the right is the courtroom. This is a well preserved Georgian court interior, marred only by the insertion of an office in the former public gallery at the rear of the room. Simple elegant Georgian joinery divides the room into the usual compartments – Grand and Petty Jury Boxes, Witness Box, Defendant's Box, Bailiff's Box, Prisoner's Box (with stairs down to the cells). These are arranged around the well of the court, which is completely enclosed, like a bloated box pew; entry is by little panelled doors. The clerk's seat is enriched with Regency mouldings; behind it rises the bench, typically with a curved back wall. The only real extravagance in the room is the back board behind the judges seat, a delightfully naive spray of Battey Langley Gothick.

A second courtroom was added to the rear of the building in a modest L-shaped ashlar extension which is well hidden. This has been converted by the new owners of the building, who appropriately are the Dean and Chapter of Ripon Minster, to provide new cathedral offices. The Georgian court has been leased to the City's Police Prison Museum, which opens it by appointment to the public. Unsympathetic modern partitions have been removed and the possibility of opening up the cells which lie beneath the building – the dock has long been sealed from these – is being investigated.

Ripon is now blessed with three museums related to themes of law and order. As well as the Court House, there is the Workhouse Museum in the nineteenth century Ripon Union Workhouse, and the Prison and Police Museum in the city's Victorian police station and gaol. The public can also visit the Old Courthouse – the remains of the medieval Liberty Court House which later became a gaol and then a debtor's prison and is now an antique shop. Few, if any, towns or cities can boast a comparable set of publicly-accessible sites and they should be jointly promoted and imaginatively interpreted in order to make the most of this happy situation.

St Albans Town Hall and Courthouse Market Place

Grade II*

George Smith's Town Hall, built in 1829-33, overlooks St Albans market place with a haughty stucco façade. The height of its pedimented Ionic portico, raised on a tall pedestal, is accentuated by the constrained width of the building. Within, the assembly room runs across the front on the piano noble, connected to the entrance hall (now tourist information office) beneath by a staircase behind, apparently rebuilt in the late nineteenth century. The staircase hall also gives access to the courtroom at the rear of the building, designed for the quarter sessions. This is big and, unusually, octagonal (although in fact the back is cut flat), rising to a large lantern, which provides the only light. Decoration is sparse, and with the white walls and ceiling and simple grained woodwork, the room feels cold and austere. But it is perfectly preserved, with little evidence of reordering. The dock faces the bench across the well of the court, and cuts into the public seating behind. The two are separated by spindly but lethal looking spikes. Steps lead down to cells in the basement. Doors behind the bench lead to the judge's and jury's retiring rooms.

New courtrooms were provided in the town's Civic Centre, which opened in 1966, and Smith's court, barring the odd book fair and filming assignment, has remained unused ever since then. There appears to be little pressure to make greater use of the room, though opening it to the public could be achieved at relatively little cost. It should certainly not be altered because, though not one of the more richly finished Georgian courtrooms in the country, it is certainly one of the best preserved. Its very sparseness seems to capture something of the harshness of the Georgian judicial process that is lost in grander interiors.

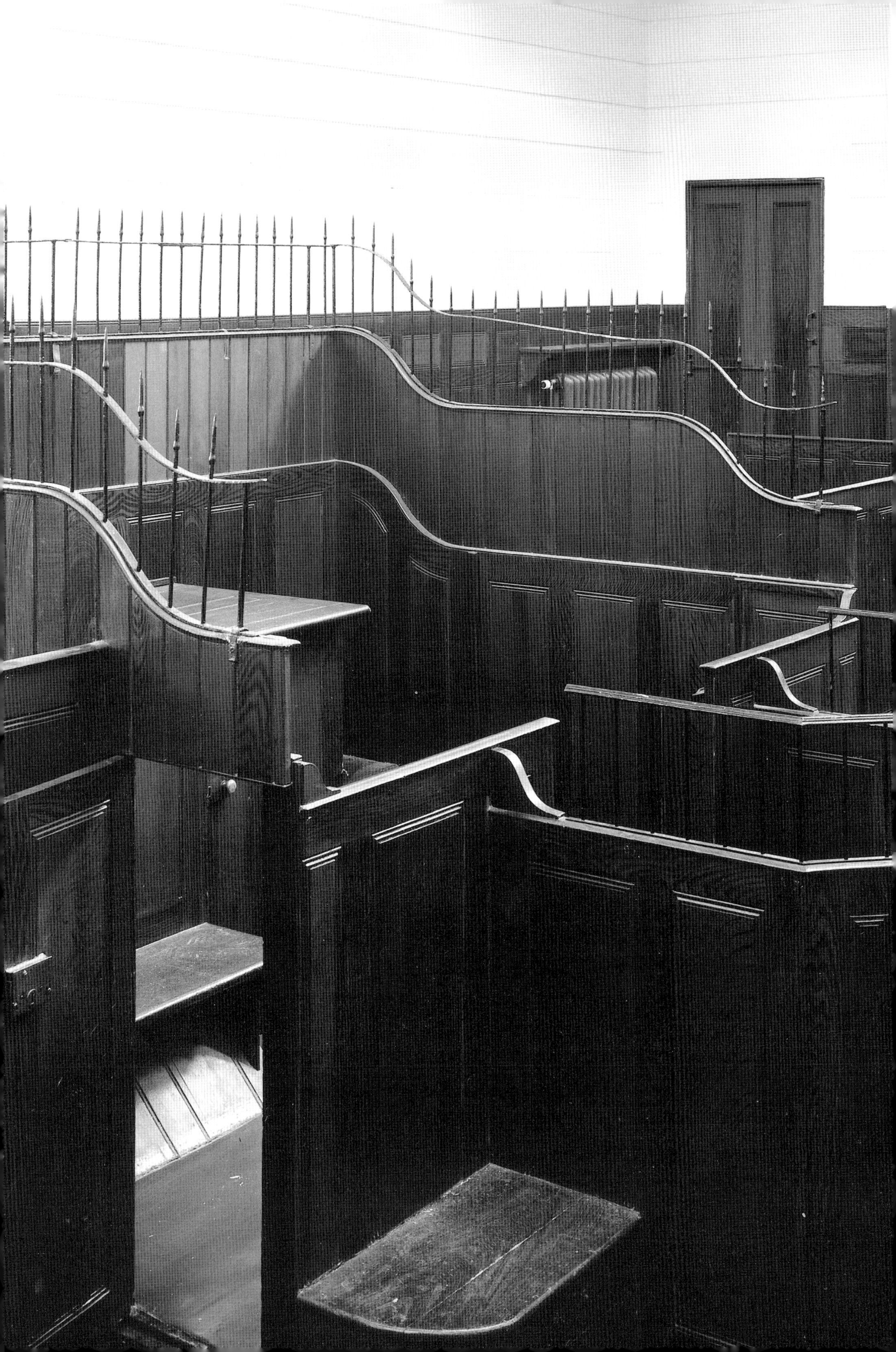

Salford County Court
Encombe Place   Grade II

Salford County Court, dating from c.1865, was designed by Thomas Sorby who succeeded his boss, Charles Reeves, as Surveyor of County Courts two years earlier. It was altered in 1928 and 1985. Sorby continued Reeves' Italianate style, though his buildings were if anything grander and Salford, with its thickly pedimented first floor windows, attic storey and modillioned eaves cornice, has more of a palazzo about it than a Reeves building. Overall, an economical but handsome design.

It closed in 1992 and has since been converted into sixteen flats. This high density arrangement has destroyed any internal sense of the building's original function (the courtroom has been subdivided), but externally the impact is minimal and the result is a success. This is important because it faces a short early nineteenth century terrace across Encombe Place, which is terminated by Sir Robert Smirke's impressive St Philip's Church of 1823. Together these form a pleasant ensemble in a wasteland of noddy-box housing and semi-derelict industrial sites.

Salford Magistrates' Court
Bexley Square   Grade II

A little down Chapel Lane from St Philip's Church and the County Court is Salford's Magistrates' Court, closing Bexley Square with a handsome Greek Revival façade and attached Doric pediment. The square itself, with early nineteenth century terraces facing one another across, is another elegant surprise in Salford's often bleak industrial townscape. The building started life in 1827 as the town hall, housing the petty and quarter sessions as well as a covered market, council chamber and assembly hall, and was subsequently extended in the 1840s, 50s and 60s and again in 1912-14. It remained the town hall until 1974. It extends a long way back in two parallel wings, one set to the left of the original frontage. Internally it has been much altered and bombed and only Court 1 remains in largely its Victorian state.

For a number of years the building has been under threat of closure. The intention was originally to build a new courthouse in Swinton alongside the post-war civic centre, but this was later abandoned in favour of transferring all Salford business to the new PFI-funded central Manchester Magistrates' Court due to open in a few years time. This plan was fought doggedly by local campaigners: Bexley Square lies in a depressed area of Salford and the loss of the court and the economic activity it sustains – nearby solicitor's offices would undoubtedly up sticks for the new courthouse – would be a blow that would be compounded by the leaving behind of a large empty historic building in need of a new use. What signal would this send to a community in need of sustained investment?

But the years of campaigning have now paid off: the Lord Chancellor's Department recently announced that the magistrates will not be transferred to Manchester after all. This may not be enough to save Bexley Square, though, if the old proposal for a new courthouse at Swinton is revived. This should be resisted as strongly as the move to Manchester was, because it would still create the substantial problem of finding a new use for the old town hall. This is a big building with lots of vacant space and few interior spaces of great quality. With will and imagination it would certainly be possible to transform it into an attractive and fully-modernized court complex.

Salisbury Guildhall

Market Place   Grade II*

This stately building commanding the Market Place was designed by Sir Robert Taylor, paid for by the Earl of Radnor and built in 1788-95. It replaced the sixteenth century Council House, which burnt down in 1780 after a banquet. Its stone is a steely grey, and the rustication wild. The Doric portico was originally recessed, but altered in 1828 (by Thomas Hopper) when a new Grand Jury Room was added above. Beneath, beyond a Doric screen, is the top-lit hall, the heart of the building, with a delicate nineteenth century iron staircase. Off to the left is the Banqueting Room, vast but restrained. To the right is the Crown Court. This is remarkably plain and nothing quite adds up, the result of drastic alterations in 1828 and further changes in 1897; originally both the Assize courtrooms were situated here, back-to-back. Although re-ordered, most of the fittings probably date from 1828. Hopper moved the Nisi Prius court to the south side of the building. Towards the end of the century, fears sparked by judicial complaints – that the Assizes would be lost to Trowbridge triggered further improvements: in 1889 a cell wing was built against the West façade in place of the original portico; and in 1897 the Nisi Prius court took on its present form, planned unusually, possibly uniquely, on the diagonal, so that the well of the court and the benches behind and above in the public gallery (supported on iron columns) are curved about a diagonal axis facing the bench and judge's seat in the south-west corner. This arrangement was presumably adopted to maximize capacity without having to breach Hopper's walls. It is complete today with all its rich oak furnishings (including an elaborate canopy over the judge's seat), from which it derives its name, the Oak Court.

Since the Crown Court moved out of the building it has continued to function as a Magistrates' Court. However, in recent years criticism about the cramped and inadequate facilities has been growing and custody cases have been transferred elsewhere. Because the Guildhall is on an island site it would be impossible to extend the building, even if it could be done sensitively. Therefore it seems certain that the Magistrates will vacate the building; when and where will be determined by a county-wide PFI scheme. It is possible they will leave Salisbury altogether.

The rest of the building, including the Grand Jury Room, is already vigorously run by the District Council as a venue for conferences, seminars, weddings, dinners and other functions. The Council is very keen to expand these activities into the court accommodation once the magistrates move out, as this occupies half the building. That will bring the inevitable clash over the courtrooms, in which the compromise is likely to be the stripping of the Crown Court in exchange for the conservation of the Oak Court. Subject to recording, and preservation of the bench, this would be acceptable: there are far better preserved and architecturally important Georgian Assize courtrooms elsewhere.

Sheffield County Court

Bank Street Grade II

A typical, but nonetheless elegant, exercise in Charles Reeves' Italianate house style dating from 1854, and extended in 1908 by Hawks. The use of high quality ashlar, with much rustication, elevates this design above many others. The building is also significant because it was Reeves' pioneer two storey model. In this design the courtroom was located above the offices, rather than behind, thus reducing the size of the site required in crowded and expensive urban centres. The building was declared redundant in 1995 and has been empty ever since. However, a scheme is now underway to adapt it for office and student accommodation, described by the Ancient Monuments Society as a 'model of its kind'.

Sheffield Courthouse
Waingate   Grade II

The Combined Court Centre

The piecemeal, evolutionary history of this building helps explain why its facades do not quite add up. It began in 1808 as the Town Hall, designed by Watson of Wakefield in 1808, with rooms for the Quarter Sessions. It was extended first in 1832-3 by W Hurst to accommodate police offices, then again in 1895-7 by Flockton, Gibbs & Flockton as a courthouse, and once more in 1908. It is built in ashlar, with rusticated lower floors. The capped clock tower is too thin to be considered a great success.

The building ended up housing the Crown Court until that moved to the new and quite monstrous Combined Court Centre nearby in 1995. Since then it has stood empty, whilst a new use has been sought. The most recent proposal was for a 1,200 capacity nightclub, but nightclubs have often proved to be only short-term solutions as they have a habit of going out of fashion and going bust, and should be treated with great caution. Whatever use is eventually adopted, the key internal features are Courts 1 and 2, late nineteenth century in oak, and the lobby and staircase. Much of the rest suffered war damage and would stand up to quite radical alteration.

Finding a new use is becoming increasingly important, because the condition of the building is now causing concern. Inadequate maintenance, the devil of so many empty buildings, has allowed the roofs to deteriorate: water has been pouring in, plaster has collapsed and dry rot is rampant. Until a sustainable alternative use for the Courthouse is found it will remain very much a building a risk, and in grave danger of further deterioration.

South Molton Guildhall
Broad Street   Grade I

Until recently this was one of the most charming law courts to remain in use, but in 2000 the magistrates sat for the last time in what is known as the Constable's Room, overlooking the centre of this quiet market town, which once grew fat on the woollen industry.

The Guildhall was built to the designs of a Mr. Cullen in 1739-43 to replace an earlier building. The pedimented façade topped with a cupola and resting on three rusticated arches fronts a building constructed using stonework salvaged from Stow, the Earl of Bath's late seventeenth century house in Cornwall, which had recently been demolished. The building is an example of the market hall type of town hall, in which the public room(s) were arranged on the first floor over an arcaded ground floor market. South Molton market is now held next door in the 1863 Pannier Market.

Inside a good, broad original staircase leads to the first floor, where a series of rooms are sumptuously panelled with more salvaged materials from Stow. They are an unexpected surprise, the impact only slightly weakened by the lurid colour scheme and some 1950s vintage light fittings. The courtroom has bolection-moulded panelling, an acanthus cornice and a coved ceiling. There are good eighteenth century court fittings, including a portable dock, witness box and bar, which can be removed when the court is not sitting. The royal arms were carved by William Puckeridge of London in 1743.

Across the landing is the intimate Mayor's Parlour, with even more elaborate panelling, pedimented doorcases, a lavish fireplace and overmantle and overdoors painted with rustic scenes (all from Stow). Beyond lies a later (1772) Assembly Room with more good plasterwork.

Devon is a large, sparsely populated county and journeys to major centres can be long and difficult, particularly by public transport. However, the efficiency demands of modern government seem relentless and now that the 'archaic' practice of magistrates sitting once a week in South Molton has succumbed to the accountants and security auditors all court users have to travel to the Magistrates' Court in Barnstaple. Is this really improving the service to the public?

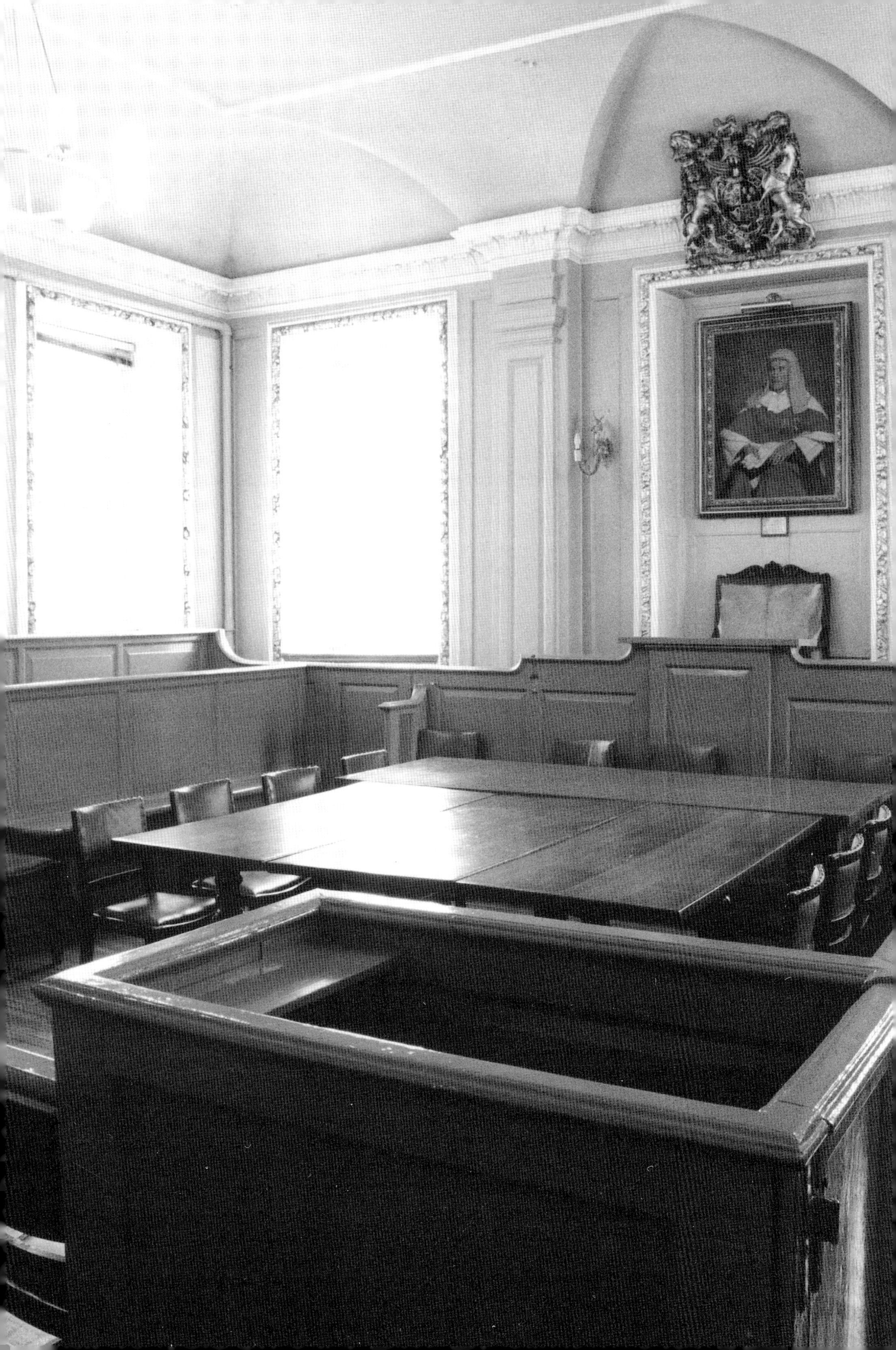

Spalding Sessions House
Sheepmarket   Grade II*

The near identical twin of Boston Session's House (q.v.), by Charles Kirk and completed in 1843. The principal façade differs only slightly: the tower parapets lack Boston's cusped panels and the ashlar is a duller local stone (the other elevations are pale brick). Inside the plan is almost identical too and Court 1 retains its original fittings. Though the magistrates' retiring room lacks Boston's wainscoting it does boast equally good furniture, including a remarkable extending fire-screen, and the magistrates' stair is certainly superior, under a vaulted ceiling with delightful cast-iron Gothic balusters. The Grand Jury Room is now Court 2. The cells remain under the courtroom with their original iron doors. In a yard at the rear is the Grade II listed Police Station of 1857, currently empty.

Spalding remains in use as a Magistrates' Court and is receiving a similar programme of repairs to Boston.

Spilsby Sessions House
Church Street   Grade II

Partly hidden behind mature trees on a quiet road on the edge of the small market town of Spilsby is a magnificent Greek Revival portico of massive Doric columns, which looks quite out of place here in deepest rural Lincolnshire. It belongs to the Sessions House designed by Henry Edward Kendall and erected between 1824 and 1826 along with a House of Correction behind it. It was considered enough of a model of its kind to be illustrated in a contemporary architectural manual.

The prison closed as long ago as 1876 and was demolished, but the Sessions House soldiered on as a Magistrates' Court until the 1980s, after which it was bought by a sound effects technician and converted into a theatre. Most of the interior furnishings have been lost, but a wonderfully rickety auditorium has been created in the courtroom, with deep blue walls and old cinema seats (see photograph, p.33). In the basement the cells retain their iron doors, but now serve as surprisingly comfortable dressing rooms; the magistrates' room has become a crush bar. Twenty years since its reincarnation, the Spilsby Theatre is thriving in the hands of the Dandelion Trust, a rather mysterious charity supporting a startling array of good works, which puts on an equally eclectic programme of dance, drama, music. In its hands this civic monument remains a central part of Spilsby's public life.

Wakefield Court House

Wood Street Grade II*

This is a significant building in Wakefield because when it was completed in 1810, it was the first major nineteenth century civic building in the town and set the scale and high architectural standard for the rest of Wood Street. Gradually, with the construction of the Mechanics Institute, the Town Hall, County Hall and County Police Headquarters, Wood Street evolved into one of the best, but least known civic quarters in England. The Courthouse is a generally elegant Greek Revival design, spoiled by the gapped-tooth appearance of the widely spaced tetrastyle Doric portico, surmounted by a figure of Justice. The ashlar is strongly etched with banded rustication. In 1849 the building was extended to the South to house a second courtroom in a sympathetic style that exactly matches the detailing of the original. The principal courtroom, later No. 1, has Grecian plasterwork and oak panelling and furnishings including a rear screen and public gallery, much altered but likely at least in part to be original.

From the outset the building housed the assizes, and subsequently the Crown Court, until the latter relocated to Bradford and Leeds in 1993. It stood empty for the rest of the decade, but following a recent sale, work has begun to convert it into exhibition space and wine bar. The wine bar is being created in the basement, and many of the cells will become snugs, an intimate arrangement that has worked well in the basement of Warminster's redundant Neo-Elizabethan Magistrates' Court (a firm of solicitors occupy the upper floors). The exhibition space will be created in Court 1 by sacrificing all the furnishings. This is always an unwelcome move, but possibly justifiable here because, unlike say Beverley (q.v.), most of the furnishings are not original. Some of those that are removed will be retained in new locations in the building. The justification offered is that in return public access and a positive cultural use for the building hopefully has been secured.

However, the scheme has closed off one possible means of modernising the facilities at the Grade II nineteenth century Magistrates' Courts immediately behind the Courthouse, which have long been considered too small and outdated. One option would have been to expand across the little back street into the empty 1810 building. Instead, the likelihood of the Magistrates' Court departing for a new building outside the city centre is now overwhelming, leaving behind another empty listed building in the heart of the city's civic quarter.

Walsall County Court
Lichfield Street   Grade II

This noble Greek Revival building, with massive Doric portico, was erected in 1831 as the Walsall Literary and Philosophical Society. The County Court offices moved in 1855, followed by the court itself in the 1860s. The red sandstone 1869 block by T S Sorby is a remarkably sensitive addition for its date – the detailing is Neoclassical and the entablature and roof line match those of the 1830s building exactly – to house the courtroom (on the first floor). Now it is the function room of 'The Old Court House' pub and bar, which opened in the building after the County Court moved out in the mid-1990s.

There is a similar story in nearby Wolverhampton, where the 1813-29 Library and Assembly Rooms was converted into the County Court. The elegant building, with a portico of superimposed columns, was sold by the Lord Chancellor's Department in 1989 and has become the 'Chambers' wine bar.

Warrington County Court
Palmyra Square   Grade II

To the left the gymnasium of the neighbouring Edwardian Technical School has been integrated, creating the biggest of the performance spaces. The others are in the Tanner building, formed in the two courtrooms on the second floor and by enclosing the old lightwell and putting in moving screens in place of its walls at ground floor level, so ingeniously enabling the newly created space to be combined with adjacent rooms.

One of the architects' trickier tasks was resolving the different floors levels (as much as a metre in difference), but the unexpected bridges spanning voids, the views opened up through glass walls, and the constant changes of level, all create spatial stimulation. At the rear, an overhanging translucent wall wraps up the new work.

The Pyramid is a key element of the town's 'Cultural Quarter' centred on pleasant Palmyra Square, which will also include the central library and town museum. Here is an example of a local authority making good use of Lottery largesse to create a new public use for an attractive, redundant civic building.

Another closed County Court, but with a refreshingly civic future. In 2002 it reopened as the Pyramid Arts Centre, following a multi-million pound conversion scheme made possible by Lottery funding. The architects, Studio BAAD of Leeds, have created a series of performance and exhibition spaces behind Sir Henry Tanner's crisp asymmetrical brick and terracotta façade of 1897. Pevsner described the style accurately as 'Frenchy Gothic'. The building also housed Inland Revenue offices.

Studio BAAD's scheme is robust, and all the better for it. A straightforwardly contemporary, full height and fully-glazed atrium, with a giant scissor brace, unites the different elements of the centre, creating a new entrance.

Warwick Shire Hall
Northgate Street   Grade I

'A remarkable job', said Pevsner. Sanderson Miller's Shire Hall (1754-8), with its giant Corinthian order and pediment, stands alongside the grave Greek Revival County Gaol (Thomas Johnson 1777-83) on Northgate Street, creating one of the finest groups of eighteenth century civic buildings in the country. Indeed Northgate Street itself is one of England's best streets, terminated by the eighteenth century tower of St Mary's Church.

In plan Miller's building follows the trend established at the Worcester Guildhall in the 1720s (now heavily altered) of moving the Assize courts out of a large hall and into separate rooms behind it. One of the glories of Warwick is that these three spaces remain largely intact, although the courtrooms have subsequently been screened off when originally they were open to the hall. This hall, 93ft long behind the facade, is lined with pilasters, thick swags and topped by a coved ceiling. But the courtrooms are the more exquisite: intimate octagons formed of free-standing Corinthian columns supporting top-lit octagonal domes, rich in stucco decoration. Iron-balustraded galleries run between the columns, and down below most of the eighteenth century oak furnishings survive, including the judge's seat and canopy. There are no prettier courtrooms in England, and somewhat remarkably they remain in use by the Crown Court. This has been made possible by the construction of a Combined Court Centre on the opposite side of Northgate, and the Lord Chancellor's Department should be applauded for both keeping the Shire Hall in use and the courts right in the town centre.

Warwick has another eighteenth century courthouse, designed by Francis Smith and built in Jury Street in 1725-8. In a niche above the door of the rusticated ashlar façade is a statue of Justice by Thomas Stayner (1731). The ground floor courtroom is no longer used by the legal system – it is now a council chamber – and few fittings survive.

Watford County Court

Lady's Close

Grade II

No one could claim Watford County Court was a great work of architecture. It is in some respects a unsatisfactory jumble of brick facades, erected at the end of the 1850s. There are some pleasant details though, such as the curious pilasters framing the main entrance, though this itself is uncomfortably tucked around the corner and awkwardly set back. Behind the main facade lay offices; to the rear, parallel to the entrance, was the courtroom. It was done no favours when Watford's inner ring road was built behind it, severing it from the town centre.

The court moved out a number of years ago, since when the building has been adapted for a new use. This has involved drastic internal alterations, carried out pretty crudely – the insertion of a new floor into the courtroom and knocking through of this new upper room to the old juror's room on the first floor of the front block. But the interior was of only minor interest, having been modernized in 1965, and it has enabled the building to find a new life as a thriving Sikh Community Centre. The new upper room is a prayer hall, bathed in warm light in late afternoon. Another welcome conversion to a community use.

Wigton Magistrates' Court
Station Road, Wigton unlisted

Wigton is another casualty of the Cumbria MCC's 'Towards the Millennium, the future delivery of justice' programme (read for that: rationalization). Presumably built as a Police Court after the turn of the last century, it is simple and unpretentious, but well built and nicely detailed. The local red sandstone ashlar is enriched with shouldered architraves around the courtroom window and the public entrance. It stands at the back of a yard (now car park), flanked by the police station, between the town centre and the railway station.

The courtroom has no fittings of great significance, but its innate virtues of natural light and generosity of proportion are being put to good use again as the setting for wedding ceremonies, now that it is has become the town's new registrar's office.

Windermere Magistrates' Court
Lake Road												unlisted

Windermere Magistrates' Court is a good example of the sort of well-built and carefully designed integrated police station and police court that were once very common and which are now rapidly disappearing (see also Chipping Campden, q.v.).

Set back from the road on a shallow terrace it is instantly recognizable as a public building. What is not as immediately apparent is quite how thoughtfully composed it is. Formal, but not too severe, the symmetrical composition carefully balances four principal masses of different heights and depths. Left and right are the police station and station superintendent's house, matching five-bay two-storey pavilions. Set forward between the two is the lower, one-storey court entrance block, the central bay enriched in red sandstone by over-sized voussoirs around the doorhead and a stepped parapet with royal arms carved in relief. Behind this rises the roof of the courtroom itself, crowned by a simple cupola, higher and more steeply pitched to identify it as the heart of the building. Materials, of course, are slate.

Inside the attention to detail continues. Across the front runs a corridor, a pleasant vista simply articulated by repeating arches. Immediately ahead lies the courtroom. The dominant feature in this generous, calm space is the deep barrel-vaulted roof. Light comes from thermal windows set on three sides and two skylights, with small leaded panes. The furniture is simple; the one exception is the monumental judge's chair.

So what happens now the court has closed? Because of the modesty of the fittings, the courtroom could reasonably be cleared for a new use, though subdivision cannot be tolerated. Everything, however, hangs on the outcome of a review by the police, who occupy rooms on three sides, of their own accommodation requirements at the site. They too are under constant financial pressure to rationalize facilities and close stations. Only then will it emerge if the police are to make use of the court accommodation, or whether it will be possible to release sufficient usable space with separate access, sealed off from the police station, for the court to be converted to independent commercial or community use.

York Assize Court

Castle Precinct Grade I

York Crown Court is far from the Lord Chancellor's Department's idea of an ideal court 'facility'. It is a single-purpose courthouse, listed Grade I, with only two courtrooms and, in many respects, inadequate facilities. Yet it seems likely that it will remain an operational court for the foreseeable future. And rightly so, for York is one of the very finest courthouses in the country, standing on one of the most historic sites.

It was designed by John Carr as the Assize Courts and erected in 1773-7, one of three monumental civic buildings arranged around a square on the site of the castle bailey. Together with the Female Prison (now Castle Museum) of

1780 it forms a matching pair flanking the County Gaol of 1701-5. On the fourth side stands the keep, Clifford's Tower. Until 1935 the whole castle precinct was a prison, enclosed by high Victorian walls. The present untidy car park was the site of the Victorian male prison.

The courthouse has as impressive a front as any in the country, though there is something old fashioned about it for its date. The dominant Ionic portico is, by English standards, loaded with legal iconography: in the pediment are crossed fasces and staff bound with laurel, and crowning it is Justice (with scales and spear) flanked by a lion and unicorn. Entry through the portico leads into a smallish double height hall, now cramped by the stagey screen cum stair inserted after 1835. The courts are to left and right. As built they were open to the hall, but they were subsequently screened off, partially so with the erection of galleries in 1812 and completely by wooden partitions between the arches in the twentieth century. Although both courts have been altered in many phases, they still retain original woodwork and much that is sympathetically nineteenth century. To make an aesthetic distinction, the seating and benches in Crown Court to the left are set out around a square, whilst that in the Nisi Prius to the right (now Court 2) radiate from a semi-circular table. Both courts are enclosed in twelve marbled Corinthian columns, and above these rise magnificent enriched plaster domes, lit by glazed lanterns. The overall effect is one of both grandeur and intimacy. A number of extensions made to the rear in the nineteenth century house the High Sheriff's Luncheon Room, a County Committee Room and other impressive apartments.

In 1990 the PSA completed a £2.8m refurbishment of the building, which included redecoration, restoration of the plasterwork and reinstatement of lost features, but which was primarily concerned with stabilizing the front wall, which for at least a century had been sliding into the infilled castle courtyard. This work should see the structure good for continued court use for many years to come, fulfilling an important civic role in the City. If it was closed, the nearest Crown Courts would be many dozens of miles ways, in Leeds, Wakefield, Hull and Middlesbrough, something that the citizens of the city would be unlikely to stand for.

SCOTLAND

Dumbarton Sheriff Court
Church Street   Category B

This was almost the one current exception to the Scottish Court Service's policy of retaining and modernizing historic Sheriff Courts.

The earliest phase of the ashlar façade is the elegant pilastered central section, James Gillespie and Robert Scott's County Buildings of 1824. The lower wings either side were added by William Spence in 1873. Later in the century police offices and a council chamber followed. Behind was a gaol, demolished in 1973. Within the oldest part is the original courtroom, now Court 1. It has good original details – the cornice with Anthemion and Palmette frieze, pedimented door cases and a delicate cast-iron balustrade to the narrow gallery. The remaining rich mahogany court furnishings are huddled around a D-shaped table below the bench at one end of the room. Court 2 was created out a council chamber. The court fittings are recent, but the room has some good plasterwork, exposed timber trusses and very good quality panelled wainscoting. The other public spaces, including the staircase, were clearly altered in the late nineteenth century. Perhaps the most interesting features of all are the Art Nouveau touches, surely the influence of Mackinstosh and the Glasgow School: a stained glass roof light and, particularly, the superb gates in the forecourt.

The building had for years required a thorough overhaul: facilities were inadequate and the roof leaked. It was slowly sliding into the burn flowing behind it, causing severe structural problems. Therefore the SCS had intended to abandon it and build a new Sheriff Court, but it struggled to find a suitable new site and in a very welcome change of heart decided to repair, refurbish and extend the existing building instead. The work is currently underway.

Dundee Sheriff Court

West Bell Street

Category B

The Scottish Court Service has an excellent track record for creating modern court buildings, with additional courtrooms and all the security and IT features and support accommodation and comfort expected, without abandoning historic courthouse in town and city centre locations. Paisley, Hamilton, Aberdeen and most recently Ayr Sheriff Courts have all received such treatment, as well as the High Court in Glasgow's Saltmarket. Dundee Sheriff Court is another excellent example of this policy.

Although the competition for a new courthouse and gaol for Dundee was won in 1833 by George Angus, only the gaol and the police office were erected under him. Construction of the courthouse was delayed until 1863 and undertaken by William Scott, Angus' pupil and the Town Architect. The impressive façade with massive pedimented Tuscan portico is Angus's, but Scott deepened the plan to create a rectangular courtroom in place of the D plan one intended by his master.

By the early 1990s the main façade and the courtroom were the only significant elements to survive alteration: the prison was demolished in 1930, and is now the site of the police station; most of the west wing was demolished for road widening; and the east is occupied by the Burgh Court. The Court Service wanted to modernize the court and expand it to five courtrooms. Nicol Russell Studios were appointed to lead the refurbishment, which was completed in 1996. This involved the usual, and tricky, tasks of providing adequate support accommodation and securely separate circulation. The extra space required was created by closing off and infilling yards to the east and west of the original courtroom and raising a new but discreet roof above the parapet. A pleasant public concourse was inserted left and right of the entrance hall along the width of the front facade, opening up into a two storey atrium at the eastern end. Behind four new courtrooms have been created, two either side of the original. Two of these are equipped with docks and direct stair access to the cells below for criminal business, and two without for civil and appeal cases. The impressive, richly decorated panelled ceiling of the 1863 courtroom has been restored. Below the high-set, tall windows, the veneered court furnishings are post-war; the seating is new.

Dundee Sheriff Court is just one example of a coherent Scottish Court Service policy to update and extend Sheriff Courts on existing, historic sites. Often this is done with the confidence to avoid historicist extensions. Always it ensures justice remains in the heart of towns and cities and that major civic buildings have a working future.

The Supreme Courts of Scotland, Edinburgh
Parliament Square

Category A

The Scottish Supreme Courts occupy an incredibly complex and historically and architecturally important group of buildings on Parliament Square, behind the High Kirk of St Giles, off the Royal Mile. The site has always been the home to the parliament and law courts which originally occupied the residencies of the prebendaries of St Giles. Today, the heart of the complex is Parliament Hall, built in 1631-40, where the Scottish Parliament met under its remarkable hammerbeam roof until 1707. The elegant classical façade to Parliament Square in the manner of Adam is by Robert Reid, 1807-34, and encases, to the west of the Hall, the magnificent Neoclassical Libraries to the Signet and the Faculty of Advocates and, to the east, the Exchequer Chambers (rebuilt twice since 1700) and a series of courtrooms, supplementing others constructed a few years earlier south of the Hall. Further rebuildings, extensions and alterations followed throughout the next hundred and fifty years, creating a rabbit warren with eighteen different courtrooms on a number of different levels. Many of these are little altered from their Victorian and Edwardian appearance.

By the 1990s the complex was clearly inadequate and rundown. It needed not only substantial repairs but comprehensive modernization and reorganization. Access for the disabled was very difficult, support accommodation (meeting rooms etc.) woefully lacking and secure circulation nonexistent: defendants had to be marched through public corridors under police escort past judges, witnesses and the public. After years of planning, a £105m fourteen year rolling refurbishment programme is now underway, with Law Dunbar Naismith as architects. The object is to completely overhaul the complex without at any point having to close it. 25% more floorspace will be created, though the number of courtrooms will actually be reduced – the enhanced facilities will enable more efficient and flexible use of the remaining ones so that overall capacity will be increased.

The key to the programme is the exploitation of the dramatic fall of the site to the south. In order to retain a largely level floor, as the complex developed it was built out over massive basements, housing cells and stores. Now these will be put to use to house enhanced custody facilities at the lower levels, with direct access to courtrooms, and administrative support, immediately below the ground floor. On this, the principal level, all the historic courts will be retained and restored and the circulation remodelled to create a clear and legible plan for users. Parliament Hall, with it portraits and statues describing four centuries of Scottish legal and political history, will remain the heart of the complex, the central lobby and assembly point for advocates, litigants and other court users.

This is the eleventh major building campaign on the site and it will be an immensely complex one. It will ensure that the Supreme Courts remain in the heart of the capital, at the centre of a legal quarter that also includes a new Sheriff Court on Cowgate, the District Court and the many chambers of advocates and solicitors, and it should guarantee the future of the site for many decades to come.

Forfar Sheriff Court

Market Street

Category B

Forfar Sheriff Court stands on an elevated site north of the town centre, a position it shares with the County Buildings. The courthouse was designed by James Maitland Wardrop, of Brown and Wardrop, and built in 1869-71. It replaced a Georgian building in the town centre, which was converted into offices for the Burgh. The style is Scottish Baronial with touches of English Tudor thrown in; the most memorable element is the roof, all crow-stepped gabbles, finials and tall clustered chimney stacks.

A grand stone stair leads to the principal first floor courtroom. This is a large, open, level-floored room, crowned by an impressive double hammerbeam roof. Most if not all of the nineteenth century pitch pine court joinery survives, carefully modernised to introduce more comfortable seating. All in all a very typical Victorian Sheriff Court, and handsome with it. There is a good solicitors' library too, with the original built-in bookcase.

Greenlaw Town Hall

Berwickshire   Category A

Greenlaw is a village of under 600 people deep in the Berwickshire countryside. It comes as a great surprise therefore to find a grand and sophisticated Neoclassical building proudly standing in the centre. It has low pedimented pavilions flanking a tall central section of two Ionic columns in antis. Above this rises a stone dome on a drum; behind is a large three bay hall. The building is Greenlaw Town Hall, the date is 1829 and the architect is John Cunningham.

It exists because for three hundred years Greenlaw was the county town of Berwickshire, though it is hard to imagine it walking around the place now.

The spur to the completion of such a grand building was the intense rivalry with Duns for the status of county town and with it the right to host the Sheriff Court, a rivalry finally won outright by Duns in 1903. Cunningham's building replaced a court house of 1712 which stood to the west of the church (whose tower was built as the county gaol) and its cost was met by Sir William Hume Campbell. The plan was straightforward: beyond the entrance lay a vestibule which opened up in turn to a staircase hall. Left and right in the wings was ancillary accommodation including a committee room and a kitchen. Beyond the staircase hall was the County Room, a large double height space with a pair of columns in antis at either end. Between the near set was the judges bench facing out over a simple table to rows of benches for lawyers and the public, occupying about half the room. Presumably these could be moved aside when the room was used for dances and other entertainments. The jury sat to the right of the bench. The county records were stored in the dome behind iron doors on stone shelves that still survive. A few years later an inn (now the Castle Hotel), built to provide lodgings for the judges and other lawyers, was laid out across the road, flanked by stables to create a courtyard facing the Town Hall. Together they form an unlikely bit of formal planning in rural Berwickshire.

After the Sheriff Court left for Duns in 1903 the building remained the civic heart of the village, home to dances and concerts. The County Room was apparently hit by a stray bomb in World War II and the court fittings destroyed. The present roof dates from then. Plans in the 1960s to demolish it were successfully fought off by villagers, and remarkably it became an indoor swimming pool, before rising fuel costs closed it. In the 1980s it was an antiques shop, but for a decade now it has been empty and decaying, and it now requires substantial and costly repairs. A local trust established to restore it has acquired it and is now actively trying to raise the necessary funds. One possibility is that it might become home to a collection of theatre organs, Wurlitzers and the like, already based in Greenlaw.

The Town Hall is one of the finest Georgian buildings in southern Scotland, in one of the most unlikely locations; it deserves all the support it can get.

Perth Sheriff Court
Tay Street   Category A

Perth Sheriff Court occupies a splendid location on the banks of the River Tay. The building was planned by Robert Reid in 1812, when the gaol behind (now demolished) was built, but the Sheriff Court itself was not begun until 1819, and then under Sir Robert Smirke. It cost £22,000. The spreading ashlar façade is dominated by a severe octastyle Greek Doric portico. Behind this is the entrance hall, with a staircase striking for its massive solid Arbroath stone balustrade. Beyond this is the principal courtroom. Now square in plan, it was originally semi-circular, and was rebuilt in its present form in 1866-7. It has altered little since then and feels, with a public gallery on cast iron columns, dense raked seating on all sides and judge's bench tight to the court, much more akin to an intimate English courtroom than many Sheriff Courts. Stairs from the dock lead to cells below. The room is lit by windows high on three sides and the ceiling is heavily ribbed with rich plaster decoration. A second court room, a bit of a dog's dinner with modern furniture, occupies the north wing, and a splendid Ballroom the south.

Stirling Sheriff Court
Viewfield Place

Category A

This is an excellent example of a Victorian Sheriff Court. The architect was Thomas Brown and it was built in 1874-76. The style is Scottish Baronial, with quite a bit of early Renaissance detail. The original façade, with its lively roof line, was symmetrical; the extension to the left in matching style dates from 1912.

The interior is dominated by the magnificent first floor courtroom and its fine hammerbeam roof. The furnishings are original, including the imposing canopy above the judge's bench and the witness box with sounding board. The walls are lined with grand portraits. Two aspects of the room strike an English visitor. The first is that, like many Victorian Sheriff Courts, it is spacious and openly planned. Nearly all the elements of the courtroom are on the same level: the floor-level public seating is only gently raked and only the bench is raised significantly; the dock is almost indiscernible, lacking high security sides, railings or screens – which makes it appear alarmingly insecure to English eyes. This is in contrast to the tight, intimate, raked auditoria of most English Assize Courts and it creates a much calmer, less intimidating and adversarial atmosphere.

The second point of interest is the Scottish Court Service's habit of gently altering or replicating original wooden furnishings to create greater levels of comfort – more leg room and padded seats – or additional seating without sweeping them away. The two issues are linked, because it is the spacious single-level planning that makes such alterations relatively straightforward.

NORTHERN IRELAND

Armagh Courthouse
The Mall   Grade A

Armagh Courthouse is an interesting building for two reasons. First, it is one of the finest in Northern Ireland. Second, as it stands today it is substantially a reconstruction of the building that was devastated by a car bomb in 1993.

The building was erected in 1809 to the designs of Francis Johnston, and was apparently his first job after being made Architect to the Board of Works. The principle façade is simple, but imposing, though the four Doric columns of the portico are oddly thin and elongated. According to Charles Brett (Court Houses and Market Houses of Ulster, 1973), Johnston was greatly distressed by this, which he blamed on the builders. But this doesn't detract from the critical importance of the building in townscape terms. It occupies a key location in the town, at the head of the great green Mall, a broad swathe of grass lined by trees and Georgian terraces.

The interior was also impressive: Brett said it was the only courthouse in Ulster with an interior to match the exterior. The plan contained the usual principal elements: two courts, hall and Grand Jury Rooms. The courts lay at either side of the hall, which was divided into two by an arch. Each half was lit by a roof lantern. The inner hall, approached from the lower by a short flight of steps, contained a fine divided staircase. At the top of this were the Grand Jury Rooms, linked to narrow galleries in the courtrooms. Following the common custom, the Civil Court was arranged on a D shaped plan and the Crown Court on a rectangular plan. There was good detailing throughout, often quite original: the public entrances to the courts from the lower hall were set in tall arches supported on elongated pilasters with oak leaf capitals; a bunch of feathers supported the cornice in the corners of the inner hall; and the Grand Jury Rooms had tent-like coved ceilings (one with arrows and scales of justice).

In September 1993 a car bomb shattered the building. The portico was very badly damaged and the interior destroyed: nothing apart from the masonry walls was capable of reuse. But the decision was taken to rebuild it, and to take the opportunity to enhance the facilities and plan at the same time. Leighton Johnston Associates were appointed architects. They restored the 1809 interiors to close to their original appearance, with an impressive quality of craftsmanship evident in the new joinery and lime plasterwork. But the exercise has raised accusations of pastiche. In fact, the building was already largely a replica as a lot of plasterwork, most of the woodwork, the staircase, balusters and roof were all replaced during repairs in the 1960s. The plans of the courtrooms were also altered during this time.

Leighton Johnston themselves do not claim that the building is an exact replica, more an adaptation to meet modern demands in the idiom of the original. For example, the new oak fittings in the courtrooms are detailed like their nineteenth century predecessors but laid out according the latest Court Service directives. And the floor area of the building has been quadrupled by excavating a full basement housing a modern custody suite (which replaced a single cell) with direct access to both courts and constructing a new extension in matching limestone and detailing in place of a 1960s wing.

Then there are the Province's security requirements. 35mm security glass was installed as the primary glazing, with fake Georgian glazing bars applied: the architects believed that this, not visible from a distance, was preferable to secondary glazing that would upset interior proportions. A bomb proof porch in the outer hall was also unavoidable. It has been dressed with pilasters and entablature. The requirement for a four metre high security barrier around the complex was met on three sides by a stone-faced concrete wall three metres tall topped by elegant railings, and in front of the building by full height railings. These are critical to the scheme. A solid wall here would have destroyed the building's fundamental relationship with the Mall. Finished with enriched gates, this is a successful solution to a difficult problem.

The Courthouse formally reopened in 1999. The £8m project raises many familiar questions about conservation, restoration and authenticity. But in Ulster these are further complicated by the omnipresent political dimension. Though for the architects the decision to restore the building was an architectural one, for some it also had a political significance. Courthouses are a symbol of Protestant, British ascendancy; not to restore Armagh to its former appearance would have been construed by some in the Loyalist community as a concession, a surrender, to terrorism.

Armagh Manor Courthouse
Ballagh, County Fermanagh   Grade B2

Brett described this building as a 'tiny, and wholly unexpected, manor court house of 1853 tucked into a remote fold of the countryside'. It is part of a series of cottage ornées built by James Haire to complement the chateau he built for himself up the hill. The rough cast building has curvy barge boards; the courtroom is signalled by the pretty little tower with a pyramidal roof.

Manor Courts were numerous in nineteenth century Ulster. Their primary function was the collection of small debts and each was presided over by a seneschal, normally the landlord's steward or agent, and a jury picked by him. Not surprisingly they were open to abuse, and the justice dispensed was not always noted for its quality. The Rev. John Groves, writing in 1819 of Ballygawley Manor Court, complained: 'The little legal information generally possessed by the person who sits as judge, and the inferior rank of those who attend as jurors, induce a probability that erroneous decisions must occur much oftener than could be wished'.

Most manor courts convened in pubs and thus purpose built courthouses are unusual, which makes the fact that this one is boarded-up and empty all the more unacceptable. It is still owned by the Haire family. There can be no excuse for failing to restore it as a family home.

Crumlin Road County Courthouse, Belfast

Grade B+

The pink and brown paint on the exterior of this imposing Neoclassical building, barricaded behind security fencing, seems strangely out of character in Belfast, as if a little bit of St Petersburg has been dropped onto the Crumlin Road. But along with the contemporary jail across the road, the courthouse is a building of great political symbolism and part of one of the largest complexes of buildings at risk in the UK.

Its origins lie in Belfast's success in the 1840s, after a century of trying, in wresting the position of assize town from Carrickfergus. That success brought with it a requirement for a new courthouse and gaol and Charles Lanyon, County Surveyor and Belfast's most influential architect, produced designs for both in 1847. Unfortunately the Grand Jury was not prepared to fund his ambitious initial design, with its majestic octastyle portico, and he prepared a more modest scheme, costing £16,500, which was erected in 1848-50 along with the gaol facing it across the Crumlin Road.

At the time it was hailed as 'one of the finest edifices of its kind in Ireland'. However, much of the character of the building today derives from the substantial alterations and additions made in 1905 by Young and Mackenzie. Lanyon's building extended six bays either side of the portico. Young and Mackenzie filled in the intermediate four bays in matching style where previously they had been recessed, and added new wings at either end, unifying everything with new fenestration and stucco. Only the portico crowned by a great statue of Justice remains unaltered from Lanyon's exterior.

Substantial alterations were also made to the interior, but at least three important spaces survive: the central hall and the Crown and Record Courts. These are substantially as Lanyon designed them, and the two courtrooms retain their original, simply

163

panelled furnishings under layers of rather sickly paint. Their layout is typical for the date, with the usual compartments raked around the well of the court, most enclosed with the box pew style doors. Public galleries rise behind a transverse gangway, the symbolic bar, that divide the courtrooms in two.

Towards the end of the twentieth century, wear and tear and the results of under-investment were exacerbated by the bomb damage the building received during the Troubles. It was a prime target, as the scene of many notorious trials, including the Diplock nonjury trials, and the prison housed Republicans rounded up under Internment in the early 1970s, and later terrorist prisoners. By the end of its working life in the 1990s the courthouse was in such bad condition that scaffolding was erected within it to protect users from plasterwork falling from the ceiling. Complete restoration and modernization was considered in 1994, but rejected because the facility it would create was judged not good enough to justify the cost. A new courts complex would be built instead at Laganside. This was funded by PFI, designed by the Hurdrolland Partnership and opened in 2002. Even the cost of keeping Crumlin Road open just until the new centre opened was considered too great and business was transferred to other courts in 1998.

All this has left a huge crumbling headache on the Crumlin Road. The PFI contract transfers the responsibility for finding a new use for the redundant building to the service provider, Dunloe Ewart, but it will find the task far from easy. Aside from the condition and size of the building, there are the problems of its location and historical associations. One interesting proposal is to convert the gaol into a store for the Northern Ireland Record Office and restore the courthouse as the public reading rooms, utilizing the subterranean passage that once allowed prisoners to be transferred securely from one to the other to transfer documents instead. Sadly this imaginative idea appears to have foundered. How long will it before an alternative solution is finally found?

Bushmills Courthouse

Main Street

Grade B1

This decaying building was built before 1830 by Sir Francis Macnaghten. There is little about it to indicate a civic building. In external form it is entirely domestic – a substantial three-bay, three-storey Georgian terraced building – though perhaps the sizeable Doric porch might suggest a higher purpose. It was fitted out on the second floor as a Petty Sessions Court, on part of the first floor as 'a kind of bridewell' and elsewhere as lodgings. Although it is the best building in the street it has been empty and decaying without sign of rescue for many years.

Caledon Courthouse
Main Street   Grade B

This austerely classical courthouse was built about 1822, probably to a design by William Murray. The off-centre pediment is crowned by a dignified octagonal cupola. The building housed a courtroom on the first floor and a dispensary and other offices below. It also apparently housed an inn.

It has been empty and neglected for many years, an eyesore in the centre of the village. Finally, however, salvation is at hand in the form of the Caledon Regeneration Partnership, which is set to restore the building, most likely as a series of office units, as the flagship element of its regeneration programme for the village. An extension at the rear will provide a kitchen, lift and additional units. Manor Architects have been appointed and if all goes well work will start later this year. The project has been made possible by public funding – the Heritage Lottery Fund's Townscape Heritage Initiative and Northern Ireland Department community regeneration funding.

The Courthouse is a prominent building, and its possible salvation has generated a lot of local interest. Indeed, more firms have already expressed a desire to move in than there will be units available.

Hillsborough Courthouse

Grade B2

Hillsborough Courthouse is a delightful example of the combined market and court house, normally designed with a first floor courtroom over an open arcaded ground floor market hall (Antrim is another Ulster example). These gradually became less common in the eighteenth century as the requirements of the courts become more sophisticated and required dedicated single-purpose courthouses.

Hillsborough, though, doesn't fit the archetype because the courtroom is on the ground floor as well, and forms part of a second phase of development.

The earliest part, of 1760, is the central two-storey section with an arcaded ground floor market hall and a clock-faced cupola. It was designed by W. Forsyth for the Marquis of Downshire. The flanking wings with Diocletian windows were added fifty years later by James McBaine. The courtroom is in the north one of these; the south one contains the Downshire estate office. Curiously the building is constructed of two different stones, the lower half granite, the upper sandstone.

In 1959 the building was transferred from the Downshire estate to the State and a restoration programme was begun which dragged on slowly for over thirteen years. The court facilities were finally made redundant in 1987 when Hillsborough sittings were transferred to Lisburn and the building passed to the Environment and Heritage Service, which appointed the Construction Service of the Northern Ireland Department to restore the building once more.

The work, undertaken in 1996-7, won a Civic Trust award. The market hall was glazed in to become a tourist information centre, the Grand Jury Room was made into a community room and the courtroom, with its simple but elegant nineteenth century furnishings, was restored and opened with exhibits, mannequins and an audio visual presentation explaining the history of the legal system in Ireland. Carefully concealed facilities enable the room to be used as a lecture theatre as well. Restoration was complicated by water penetration and the ceiling and walls of the Grand Jury Room had to be replaced in matching fibrous mouldings and lime plaster.

The project has been judged a tremendous success by the people of Hillsborough, who have enthusiastically taken to their 'new' civic building. It is constantly busy with meetings, drama and musical productions, clubs, dances and weddings. There can be no more welcome future for a redundant public building than as a vital new centre for community life.

Londonderry Courthouse
Bishop Street   Grade B

Londonderry's Courthouse is one of the finest Neoclassical buildings in Northern Ireland. It was completed in 1817, the architect was John Bowden and it supposedly cost £30,759 (twice as much as the Belfast's larger Crumlin Road thirty years later).

The dignified tetrastyle portico is of the Ionic order from the Erechtheion on the Acropolis, with its bulging volutes, and it is embellished with symbols of state power and justice: a substantial Royal Arms above the pediment and statues of Justice and Peace above the flanking wings (the architect signed the back of the Arms). The statues were not part of the original scheme; Bowden's plans instead included a Greek inscription from the Menander for the entablature that translates as, 'This is the eye of Justice that sees all things'. The two courtrooms – Criminal and Civil – are perfect 32 ft cubes. They are located behind the entrance, or Great hall, and were originally only semi-divided from it, by broad openings framed by columns.

The interior has been much altered over the years and the courtrooms have lost many of their furnishings (though they retain their tall roof lanterns). In 1988 the building was badly damaged in terrorist bomb attacks. When the damage was repaired the opportunity was taken to completely refurbish and modernize the building. It officially reopened in 1998, equipped with four courtrooms, expanded support accommodation, wheelchair access and a comprehensive IT network, all provided at a cost of £7m. An excellent job was made of restoring the creamy external masonry and, fitted with surprisingly elegant new railings that manage to satisfy the stringent security requirements without masking the building from the street, it has reasserted itself as a noble part of Londonderry's townscape.

Middleton Market House
Main Street   Grade B2

An example of how simple some of these buildings can be, though it did originally sport the obligatory clock-face cupola. This was built in 1826 and housed a Petty Sessions Court on the first floor and an arcade market hall below. It is now in a very sorry state, having been empty for at least thirty years. Perhaps it is surprising that it is not in worse condition.